The Meaning Of Life

The Meaning Of Life

The Meaning Of Life

Copyright © 2008 by Mark Morris. All rights reserved, including the right of reproduction in whole or in part in any form.

Manufactured in the United States of America

ISBN 978-0-578-00422-8

Cover art created by Mark Morris

The Meaning Of Life

ABOUT THE AUTHOR

I am a software engineer by profession who has developed unmanned autonomous vehicles used in space, in the air, on the ground, and underwater. I'm a singer, songwriter, guitarist, drummer, producer, painter, rock-climber, and a voracious lover of life. Born and raised in New Palestine, Indiana, I received a BS in computer science and a minor in art from Purdue University. I spent seven years working in Silicon Valley California after college before settling down in Cary, North Carolina to raise a family with my wife Sasha. Before moving to North Carolina, my wife and I toured the world for three months. This trip inspired in me a deeply reflective state of observation and concentration that revealed an epiphany on the meaning of life. Before that time I had never harbored any aspiration of becoming an author, but I felt so moved by this new paradigm, calling me to look anew in awe at everything around me, that I wanted others to experience the peace I have found within it.

The Meaning Of Life

```
For my dad, who taught me the power
    and joy of asking questions
```

The Meaning Of Life
By: Mark Morris

1. Introduction To Order
2. Functional Order
3. Influential Order
4. Context
5. Order Is Winning
6. The Human Vessel Of Order
7. Level Of Order As Importance
8. Tasks
9. Finding Wisdom
10. Finding Creativity
11. Education
12. The Future Of Humanity
13. Is There A God?
14. The System Of Society
15. Groups
16. Free Will
17. Everything
18. The History Of Order

Chapter 1: Introduction to Order

This is how I plan to redefine the word "order". Order is the combination of its two parts, which I'll refer to as functional order and influential order. An understanding of this kind of order provides us with answers to questions that have perplexed human thought for thousands of years. What is the meaning of life? Why do we do the things we do? How does one achieve wisdom? Do we possess free will?

As humans, we are the most functional beings known to exist in our universe, except perhaps for a creator should he/she exist. Because I could never understand God except perhaps through his/her creation, should he/she exist, I'll focus solely on what I have come to understand.

In the beginning of our universe's existence, our matter and energy burst out into space in the most powerful of explosions. All known matter was dispersed in a disorderly cloud of dust, racing away from the central point of our universe. This matter advanced out into greater spatial distance, twisting and turning, pushing and pulling, guided only by the constant fundamental forces.

The 4 fundamental forces (gravity, electromagnetic, strong, and weak forces) have acted upon matter since the beginning of our universe by spatially affecting that matter throughout time in a constant way. Every atom in our universe has an effect on the movement of every other atom.

The strong force bound together subatomic particles at extremely small distances with an extremely powerful force creating the nuclei of our atoms. The electromagnetic force bound together protons and electrons creating atoms, and bound those atoms together creating molecules. While the electromagnetic force pushed and pulled at particles of matter, grouping attracted elements neatly together, while separating others, gravity only pulled matter together. Through the force of gravity, this massive disorderly cloud of dust grouped into more orderly clumps of matter we call stars and planets.

Throughout space and time these fundamental forces have acted upon matter by giving it higher order on various scales of distance. For example, the near perfect spherical shape our Earth has taken from gravity mimics how a water droplet takes shape through molecular forces. As time progresses the matter these forces act upon becomes more orderly, grouped into more useful, resourceful, functional bunches. Through an understanding of the nature of order we can gain powerful insights into the inner-workings of the world around us.

Any system has a definite order and a definite level of order. On the surface, order is what we all understand it to be: the grouping of similar things, things arranged in a line, notes in a particular scale, letters forming words on a page, arrangements of C, G, A, and T molecules on a strand of DNA, etc. When there exists a minimum set of material to be disorderly, a certain level of order also exists within that system. Every grouping of things, every person, idea, song, book, tree, painting, all have a definite but unknowable level of order.

"Why discuss it then?", you might ask. We don't have to understand anything's level of order perfectly to get what we want, we simply need to be better at approximating the level of order of things, and we can do that by better understanding the nature of order itself.

As humans one of the highest order functions we perform is our approximation of something's level of order. We're searching for a wiser (more functional) person to listen to, to put our trust in. Trust itself is a mechanism by which we create order adding to the influential order of the system we trust. If we find someone worthy of our trust we communicate with an openness related to our level of trust in that person. In doing so we add order in the form of knowledge to that person.

The Meaning Of Life

Perhaps that person didn't want the knowledge and will refuse to return the favor of our trust. When we trust someone we make assumptions about their order. When we interact with that someone we act based on assumptions from interactions with similar people in a similar setting and through our experience we better order our assumptions.

To begin, I should better define some of the semantics I'll be using. The word "order" has been used in many different contexts to describe many different things. In the laws of thermodynamics, for example, formulas of entropy measure order and disorder in atomic and molecular assemblies. This use of the words "order" and "disorder" describes the properties of energy relative to the absolute temperature of a system. I'm using the terms order and disorder to describe a different phenomenon. When I use the term "order" I'm referring to the arrangement of any system or grouping of matter.

When I use the term "system", I'm referring to any grouping of matter. Every system has a level of functional order and influential order that I'll describe as being of a higher or lower level. Every system has a level of order that is a combination of its level of functional order and its level of influential order. The level, high or low, of any system's order describes how resistant that system is to disorder. Before there was life or reason, a system had high functional order if it was able to resist being changed by its environment, because most change was disorderly. For life forms that can affect change through reason, the level of functional order is determined by the tasks that life form can perform and the amount of resources that must be spent to execute those tasks and receive their desired results (i.e. the efficiency of those tasks). We avoid disorder by executing our many tasks to receive higher order and by improving upon those tasks.

Imagine you have a system of friends that you like to play music with (system M), and another system of friends that you like to watch football with (system F). System M likely has a much higher functional order than system F for the task of forming a band.

Similarly if you have a system of iron particles bonded together in the shape of a spoon, that system will have a higher level of functional order for eating soup than a system of wood and nails in the form of a house. However, the house will have a higher level of functional order at providing a place to live than the spoon would have.

A system has a level of functional order at a specific task and a level of objective functional order which takes into account its total importance, only part of which relates to its functional order at any one task.

Part of what determines a system's level of functional order is that system's location. Because carriers of order must expend energy to access or experience a system of order, that system's proximity to carriers of order that can use it is a factor in determining its level of functional order.

A system of functional order can also have influential order when part or all of that system is replicated in some form. Because a system's order can depend on both its function and its influence, I will use the phrases "functional order" and "influential order". It is sometimes necessary to differentiate between these two properties of order to better understand a system's total order.

A teen pop star may have a relatively low functional order compared to the average person, while having a much higher influential order. Conversely, a reclusive genius can have a much higher functional order than the average person, with a lower influential order.

The following is a list of the major terms I'll provide as a reference to look back on when things get confusing. These terms will be explained throughout the chapters, so there's no need to understand them perfectly just yet:

Order (objective order):
An arrangement of matter

Functional Order (objective functional order):
An instance of an arrangement of matter

Influential Order (objective influential order):
All functional order created as a reflection of some functional order

Relative Order:
Received order

Level of Order:
The ability of some order to resist disorder

Level of Functional Order:
The ability of some functional order to resist disorder

Level of Influential Order:
The ability of some influential order to resist disorder

Level of Relative Order:
A being's approximation of an arrangement's level of order

Level of Relative Functional Order:
A being's approximation of an arrangement's level of functional order

Level of Relative Influential Order:
A being's approximation of an arrangement's level of influential order

Now that we have the semantics down, let's move on to describing order:

Chapter 2. Functional Order

A system of functional order is a single instance of an arrangement of matter. Once again, the level of functional order within a system is determined by that system's ability to resist disorder.

One way a system can resist disorder is to find an arrangement that is structurally sound, whereby it resists change. An example of this is a smooth stone that has no crevices or protrusions. A crack gathers water that freezes and expands, destroying the order of the stone. A protrusion causes resistance and is more easily worn away than the smooth portion of the stone, once again changing it's structure. I'll refer to this order as primitive functional order.

Another way a system can resist disorder is to improve the order of a rational life form that uses it, receiving protection from that life form. An example of this is an attractive painting that remains for decades unchanged to preserve its original power.

An even more functional way to resist disorder is to find an arrangement that resists disorderly change, but does not resist orderly change that leaves behind a system more resistant to disorder. An example of this is an idea in the curious mind of someone who is always searching for a way to improve upon it.

Like the latter two examples, any system that is used by a life form has a level of order determined by the amount of resources that must be spent by a life form to use that system and receive the desired result. These non-primitive systems include life forms and all things created by them.

Primitive functional order in the form of solid matter found higher order by finding arrangements that are more resistant to change, which created a more stable, consistent, environment around it. Before there was life or purpose, change was the result of structures colliding, and chemicals reacting with one another from the raw unintentional interactions of the fundamental forces. When structures would collide, those solid structures that were more structurally sound preserved their structure.

As gasses and liquids coalesced through fundamental forces, they formed the bodies of the atmosphere and ocean. They weathered away at the solid structures of the Earth in a somewhat predictable way. The rain always contained primarily water. That water always fell towards the center of the Earth, striking solid surfaces on the same side as long as that side faced the sky.

As primitive functional order in the forms of liquid and gaseous matter interacted with the solid matter, helping it to find more stable arrangements, the liquids and gasses of the environment became more pure and consistent.

While the weather at any given place or time is still fairly unpredictable, the Earth's interconnected weather systems follow large orderly patterns of flow that leave consistent patterns of orderly activity. Certain regions have consistent climates with predictable annual rainfall.

All of these predictabilities and consistencies create isolated regions with unique forms of primitive order. Consistency itself is a manifestation of a high level of primitive order. Consistency brings repeatable results through the collisions of matter.

Consistency and repeatability describe the higher levels of primitive order that were necessary for the formation of life. If the necessary atoms were to assemble in the arrangement of a DNA molecule in the disorderly far reaches of space, that DNA molecule could not find the atoms around it needed to form life. Consistency and repeatability gave matter the opportunity to find a structure with repeatability in a consistent environment, which eventually led to life.

Once order found life, primitive order continued its gradual increase in order while life raced ahead finding new structures at higher levels of order with unprecedented speed.

The Meaning Of Life

Life was the new paradigm order had found to reach higher levels of order.

Once life found the capacity of reason, structures were changed through reason to produce a higher order result. These structures achieved their level of order through their efficiency at producing a desired result for the beings that used them.

If system A and system B can be used to get the same result, but system A uses more resources, such as time and energy, than system B, then system B is of a higher functional order for achieving that result.

Let me provide a parable to describe the creation of higher order through human reasoning:

Imagine an old man who has lived his life in the same small city. He's sitting on a bench just outside the city limits; a place he likes to go to feed ducks in a pond. Every day he watches tourists arrive at a nearby train station and walk past him, eventually stopping at a fork in their path. The tourists then, having little knowledge of the town they're exploring, arbitrarily choose to go left or right.

The man observing the tourists understands that the divergent paths don't form a loop, so the tourists who choose the left route will have a very different experience from those who choose the right. Furthermore the man understands the left path leads to a muddy swamp whereas the right path leads to the most beautiful outlook over the city near some hotels and restaurants.

As the man walks by the fork in the trail on his way home he notices some stones strewn about on the ground. He notices how some stones are all black while others are all white, yet many are of a similar size. Through his observations he recognizes a good use for those stones. He gathers handfuls of them and builds a mosaic of an arrow pointing to the right at the fork in the path. He stands back for a while and notices the

The Meaning Of Life

tourists begin heading to the right and he smiles a smile of gratification.

The system of stones that comprise the mosaic of the arrow were given higher order by the man's arranging them. Before they were arranged as a mosaic few people would ever notice them so their disorderly arrangement was never deemed worthy of remembering. Because of this, their arrangement had little influential order. Before they were formed into a mosaic a person could look at them and extract little meaningful information, therefore they had little functional order.

    The system of stones did have a fair level of functional order before the man arranged them, but it still had a lower function and influence, i.e. a lower order, than it has as a mosaic. The stones already had similarity in size, similarity in location, and their coloring of black and white, which made them of easy form to build a mosaic.

    The man didn't have to travel far to quarry the stones, nor did he have to spend energy reshaping them. Much functional order already existed within the system of stones, enough to be easily assembled by an old man unwilling to travel far to build his mosaic. This illustrates an important concept: For higher functional order to be formed, enough order of a lower form must already exist.

    The order within the system of stones forming the mosaic first existed as an arrangement of neurons in the man's brain. The man's attention to his surroundings provided him with an observation of tourists' paths. Through processing the apparent confusion on the faces of the tourists as they pondered which way to go, with the knowledge of which way he would want to go if he were the average tourist, he found a desire within him to inform the average tourist of the better path as he saw it.

    If the man was wrong in his assessment of the average tourist's wish, say perhaps he had a fondness for muddy

The Meaning Of Life

swamps and made an arrow mosaic pointing to the left, then enough tourists would visit the muddy swamp knowing the arrow had pointed them there. The disappointed tourists would likely change the mosaic to point to the right so as not to let it fool the next tourist, or at least destroy the mosaic leaving a 50% chance a future tourist would choose the more scenic route. Either way the average person would make the system of stones more orderly or leave them alone. We all go about the world trying to increase order and through this process order increases itself.

Thomas Edison took over 1,000 tries before making the first practical (functional enough to be commercially viable) light bulb. When asked by a reporter about his 1,000 failed attempts he replied that they weren't failures, he learned 1,000 ways not to make a light bulb.

This is how our highest functional order is currently found. Edison lived at a time when man already had an understanding of electricity. He studied the current science on electricity and found a practical use for exploiting its properties. He then used trial, error, and reason to build successively better models for what would become a practical light bulb. Every model was a slightly higher form of order, first existing as a modified blueprint from the last failed attempt, in the form of an arrangement of neurons in his mind.

Edison's creation was such a functional piece of order that it took on a life of its own, gaining influential order, when everyone wanted to buy one. Before Edison emerged from his laboratory with the first practical light bulb, his idea had great functional order, but no influential order. Once it was shown to others in the public it took on great influential order.

The 1,000 failed attempts are like all the matter in the universe that doesn't support life. They are mixtures of matter in various concentrations and arrangements, where the necessary arrangement doesn't exist to create a level of functional order sufficient to spread itself with variation.

Chapter 3. Influential Order

The other part of a system's order is that system's influence. A system's influential order is all order created as a reflection of that system's order. The level of a system's influential order is the degree to which that system's functional order has spread.

A Ford Mustang, for example, has a collection of blueprints and specifications for its design that describe the Ford Mustang's order. A single instance of that design is a single car in the form of a Ford Mustang, which is a piece of functional order.

The influential order of the Ford Mustang includes all of the cars of that model that were made. It also includes the portions of all pictures taken of a Mustang that contain the car. It includes portions of all memories people and other animals have stored in their neural tissue about events that include a Mustang. Any time the design, or parts of the design, of a Mustang have been recorded, remembered, copied, or reproduced, in any medium, that action has added to the influential order of the Mustang.

Einstein's theories of relativity are a model for high functional order. Had he come up with his theories and kept them to himself, or died before he could record them in an orderly enough fashion to be understood or tested, had he been unable to pass on this magnificent functional order, the ideas would have died with their carrier purely by a lack of influential order.

The ideas, before Einstein's death, would have still had the same high functional order, but having never been passed on to another carrier of functional order, the ideas' lack of influential order would have rendered their total order useless upon his death. This example illustrates the boundary between functional and influential order, and how influential order gives the order of some functional order redundancy, which helps it to further resist disorder. When functional order is

The Meaning Of Life

transferred to another carrier of order, that functional order is given influential order. When functional order is transferred imperfectly, like through experience or communication, it takes the form of life.

The property unique to all forms of life, one might say the definition of life itself, is the ability to reproduce, with variation, in its environment. Plants, animals, bacteria, and viruses all exhibit this characteristic, as does information. When an idea with sufficient function is formed in the mind of an individual, that idea can be given life through communication.

Once that idea, or "meme" as coined by Richard Dawkins, is communicated, it follows the rules of evolution like a gene in our DNA. It copies itself with variation in the form of physical arrangements of neurons in the brains of its hosts, and thrives to the degree of its level of order. That is, you're more likely to communicate information you find useful (functional), and you're more likely to receive it if it has already spread to many people (influential).

A meme is an extremely compact and efficient piece of order. This is because order inherits the functional benefits of its ancestral systems of order. A meme, in the form of words in a language, inherits the functional order of that shared language. A language inherits functional order from the human race, like our capacity to memorize the common application of an arbitrarily chosen string of characters, and other rules such as grammar and context. In the English language, for example, this allows combinations of a set of 26 characters to express over half a million distinct words, providing a rich yet manageable system for memorization.

The human race inherits functional order from the DNA molecule, which has made evolving life possible on Earth. It is not known whether the DNA molecule was brought to Earth by debris from some other life-supporting planet, or formed on the Earth itself. Either way, the DNA molecule inherits functional order from the planet of its conception. It receives influential order from the high level of primitive order within the Earth, which provides it with an orderly environment to exist within and spread.

There are countless known fields of study that leave much to be discovered such as the intricacies of how the human genome forms our marvelously complex biology. There are even many fields of study yet to be discovered. We continually make discoveries that shed light on the ways in which order has progressed, and illuminate new ways for order to progress. We have come to these discoveries through an evolving set of memes that comprise the highest known system of functional order in the universe, all recorded human knowledge.

Again, as human beings, we are the most orderly beings known to exist in our universe, but the order we have created, in our recorded system of human knowledge, is greater than any one being's order. When we tap into that vast resource we can receive the functional order of other higher ordered beings in a much more functional way than in the past. This is an example of how order constantly finds easier ways to further itself. Often it does so through the highest beings of order (humans) finding tasks of a higher level of order (using less time and energy) to use for furthering order. To have a better understanding of order we have to stop thinking of it as something we humans have created. Order created us.

A being, or life form, is a piece of order, but order is not a being. Order has no desires. It is simply the exception to disorder. It is a property of all matter. Where enough material exists to be disorderly, order will exist within that material. Primitive order "finds" higher order like a rolling marble finds a crack in which to rest. Primitive order found the property of replication, then variation (life), which found the property of reason and used it to more selectively find higher order.

Darwin recognized the properties of life, replication and variation, to be the governing principles behind the evolution of the species. Furthermore, evolution should be

The Meaning Of Life

understood to be a level of maturity in the natural process by which matter increases its order over time. Order should be understood as the creator of life and the guiding principle behind all of life's actions and creations.

Order has created every concept imaginable, because these concepts exist as ordered matter in the form of neural networks within our brains. Order has created all of our emotions, experiences, memories, and possessions. Order has also created atoms, planets, stars, and galaxies.

All order in our universe is the product of matter, energy, time, space, and the fundamental forces. The amount of matter and time, and the number and nature of the consistent forces at play determine the possible level of order that matter can achieve. The most fundamental particles that comprise all matter are themselves not made through order, but are the building blocks used by order to make atoms, molecules, life forms, intelligent life forms, memes, and technology.

The moment an arrangement of matter first replicated itself is the moment in which functional order begat influential order. Once order copies itself it affects (influences) the other copies by competing for resources in their shared environment. This new carrier of order can persist, consuming lower ordered material, increasing the order of that material.

Huge waves of disorder can destroy houses, people, cities, even civilizations and species, but whatever order exists in its place, given enough time, will rebuild past the old order because that's what order does. Order improves itself eventually finding the property of life. Life can learn from its defeats, as long as some life remains, while disorder stays predictably disorderly.

Influential order began as nothing more than replication and then progressed into competition for resources once it found variation. One slightly different life form's ability to compete more effectively for a resource enhances its ability to replicate its order, creating influential order of that variation. As the carriers of functional order matured, so did the nature of

The Meaning Of Life

influential order. It progressed into cooperation between carriers of order where cooperation proved functional. This evolved into communication to warn others of danger or signal a coordinated attack. Today we create influential order by communicating all types of functional order.

A person's level of influential order is determined by the levels of order that person has imparted on other carriers of order. If the functional order a person offers is useful to those who receive it, they will effectively incorporate that order into their own systems adding to the influential order of that person.

In many ways the World Wide Web is just an efficient extension of our society. If we want to find ways of increasing our influential order we can study the concrete logic that defines influential order on the web. A web page exists as virtual space where functional order can exist and evolve. A web page usually belongs to a person or entity that is represented by that page, therefore that page will have a reputation, like a person, with a level of order assigned to it.

The content on a person or entity's web page is often some of the most functional content that person or entity has to provide. It is the content that a person or entity has determined might be useful to users of the web. Much of this is functional order we would offer to show only to those people we know personally, if there was no web.

One great power of the web is our capacity to submit and receive knowledge anonymously. This removes the necessity of having social skills sufficient to form the relationships previously required to contribute or extract meaningful information. This is a great catalyst for functional order's ability to find influential order. It has broken down many barriers put in place by large media conglomerates determined to control the flow and content of information.

This greater access to receive and contribute to recorded human knowledge has democratized information. It has given everyone a voice in determining what order is functional, and therefore worthy of influence. This enables

The Meaning Of Life

order to be more impartially and accurately judged on the basis of its function, with less emphasis on the current influence of that order.

The vastness of order available to us on the web created the need for highly functional search algorithms to help us find the order we're looking for. The computer logic that defines which web pages are returned from our searches reflects the ways in which we search out higher order from society.

Google maintains what it calls a "page rank" for all web pages searchable through Google. This page rank is Google's estimate of the level of a page's influential order. When someone googles a phrase, Google searches for that exact phrase, and slight variations of that phrase, in all accessible web pages, and returns a list of pages with a match. Google then puts those resulting pages in order where the top page that is seen is the page that Google's algorithms have estimated to be of the highest order relative to your search.

This estimation of a page's order for the corresponding search takes into account the page rank of the page containing the search phrase, because a person would first seek information from an influential source. If people put their trust in a person's functional order, they approximate that order to be of a high level of function. This trust they place in a person's functional order adds to its influential order. Similarly, if web page A is linked to by many other pages, the other pages are adding to page A's Google page rank. If page A's page rank (influential order) is high, and it links to page B, then page B's page ranking is greatly increased. If page A's page rank is low then it doesn't give as much of a boost to page B's page rank by linking to it.

This is very similar to how people transfer influential order. A person's influential order doesn't end with those directly influenced by that person's functional order. As a person's functional order is passed on, from person to person, the influential order of that original functional order is increased. Also, when a person receives functional order and

The Meaning Of Life

passes it on, the order that person receives grows in influential order.

Google also takes into account how closely the search phrase matches the pages that were found and how many times the phrase occurs in those pages. This enables us to reach content more functionally matched to our search even if it isn't on a highly influential page. Therefore, it is easy to gain influence on the web by offering functional information without already having influence.

Most of the order within the human mind is received order, or relative functional order, adding influential order to the original source of that functional order. Because we all interpret the functional order we receive differently, that relative functional order we've received is partially original order once we receive it.

When we come to realizations on our own, those realizations contain original functional order even if someone else has, without us knowing, reached the same conclusion. When the same order originates from multiple beings, that coincidental functional order is not influential order because it was formed through creation rather than experience.

Often our realizations come from applying some lesson we've learned (relative functional order) to our own context. Because each being's context is unique, it's possible to create functional order and pass it on to another being that finds a context for it, with a higher level of order. In these cases, relative order can take on a higher level of order than it had originally. In most cases, however, due to the limits of communication, perception, and comprehension, the order we receive is of a lower level than the original.

Chapter 4: Context

A system's level of functional order is determined not just by the arrangement of the atoms that comprise that system, but also by that system's context. A system's context is that system's position in its larger system, which is ultimately the system of all matter in our universe.

We can increase the level of functional order within a system by rearranging the atoms of that system, adding or removing atoms from that system, or by changing the context of that system in its environment (the larger system of which it is a part).

The mosaic parable provides a nice example for understanding context. If someone were to change the context of the arrow mosaic by moving it far off of the path, then it would lose its function, or take on a new function, to those that see it. This would give it a different functional order even if all of the atoms of the system were in the same arrangement in the new location.

If the arrow mosaic were simply rotated by 180 degrees, it would have the same effect as rearranging the system to point to the left, and would then become disorderly by guiding tourists to a muddy swamp. However, if someone were to rotate the arrow mosaic that points to the right by only 1 degree, clockwise or counter-clockwise, the level of the mosaic's functional order would be essentially the same. As long as the average person is guided to the right by the crude mosaic, it maintains its functional order. This example shows us how a system of order doesn't have to be perfect to achieve its orderly result. The fact that a system only needs to be sufficient to provide the same orderly result leaves an infinite number of ways in which even the most complex arrangements of order can be found.

Because a system's level of functional order depends on its location within the system of all matter, we would have

The Meaning Of Life

to understand all matter to perfectly understand any system's level of functional order. To get around the impossible task of understanding all matter we have evolved a concept I'll refer to as *isolation*. Keep in mind; we don't have to have a perfect system we use to estimate something's functional order, only a sufficient one.

Isolation is a method we employ when we define a system. Because all of the matter in the universe is a system, any grouping of matter we choose to call a system is actually just an isolated subsystem of the same system of all matter. As we try to form functional concepts to describe the system of all matter, we start out by recognizing subsystems that are objectively isolated. We also artificially isolate subsystems when their greater system is too complex to look at as a whole.

Any system that is not easily affected by changes in another system can be considered isolated. A clump of matter existing in deep space can be considered extremely isolated because there are no air currents, no moisture, and little energy from distant stars around to affect it. If a system is sufficiently isolated, we can learn about that system by making changes to the system and correlating the different results we receive from that system to the changes we've made.

No one atom in our universe is perfectly isolated from the rest because some force/forces determined by other matter are always affecting it. However, certain groupings of matter, for many different reasons, exist in varying degrees of isolation from their environment. We have developed rules in the form of neural matter to determine how to conceptually isolate systems, and how to approximate a system's degree of isolation.

While isolation is a powerful mechanism used to understand the essential concepts of a complex system, certain western schools of thought have placed too much emphasis on isolation when forming an understanding of a system. Conversely, eastern schools of thought often place too much

The Meaning Of Life

emphasis on the whole system, failing to understand its mechanics from the bottom up.

A crude example of this east vs. west mentality is in the field of pharmaceuticals. Western medicine, in general, gets the basic ingredients for its drugs from plant material by extracting one pure chemical in isolation, and throwing away the rest. Eastern medicine, on the other hand, generally relies on using whole plants (herbs) as a means of bringing order to our bodily system.

The western method might attack disorder in our body by finding a chemical supplement that improves one statistic that has been shown to identify a risk of heart attack. Without understanding how that chemical works with other chemicals to increase heart health, we sometimes do more harm than good through isolation.

On the other hand, improving our bodily system from the top down through eastern methodology involves a lot of guesswork. When we provide a root that is known to help with arthritis maybe its effectiveness is from a single chemical and we end up with a lower dose of that chemical while other chemicals in the root cause adverse side effects.

There is no right or wrong methodology, and my example is greatly simplified, but the most functional way to approach such problems will usually be found with an understanding of both a system in isolation and that system's context in its environment.

The same wisdom should be applied to our politics. Neither Democratic, nor Republican, nor Libertarian ideologies are most functional for finding a solution to most problems. On occasion the best solution lies with one party or another, but usually it's somewhere in between. We who are fortunate enough to vote can do a great service to our country by finding a candidate willing to defy their own party's philosophy at times when it proves practical.

Effective pragmatism comes both from an understanding of the current state of the variables at play, and

The Meaning Of Life

from an understanding of the rewards and pitfalls of the various methodologies one could use to find a solution. To accurately approximate the level of order of a solution to a problem, we have to understand both the system of rules that defines the possible solution, and the context in which that system will apply. The first tells us how close a solution is to ideal. The second tells us how practical a solution is.

Chapter 5: Order Is Winning

To understand how the advancement of order is responsible for everything in existence just look at any arrangement of matter and ask yourself, "Why does this matter continue to exist in its arrangement?" The answer is because that arrangement of matter has an ability to resist being changed from that arrangement. Either some being/beings with the ability to act on the environment have not changed its arrangement because it is useful to it/them and deserves protection, or the current forces acting upon all matter have been unable to change its arrangement for a period of time. When you consider other beings are a part of the arrangement's environment, then their protection of that arrangement is just another way that arrangement is able to resist being changed by its environment.

When an arrangement of matter resists change, it resists losing its arrangement, or functional order. Therefore, it resists disorder. An arrangement can be disordered through change (breaking the current order), and become part of a new system with higher order (higher ability to resist disorder). When we apply change in a somewhat random way, as it was done before reason, we only rarely create a system of higher order. This is still sufficient for the order of the system and its environment to increase, it just happens slowly. When we apply change in an orderly way (such as through reason) we usually create a system with a higher level of order, and so the order of the system and its environment increases much faster.

Arrangements with a higher resistance to disorderly change are of a higher level of order. Throughout time, our highest order materials and structures take less and less human resources to create (time, energy, material, and money), yet they take more and more of nature's resources to destroy (time, mechanical energy from wind, mechanical energy from water, chemical energy from water, thermal energy from heat/cold), while still taking reasonably little human resources to destroy,

The Meaning Of Life

making the disordering of our property increasingly under our control.

Now look at any arrangement of matter and ask yourself, "How did this arrangement of matter come into existence?" The answer falls within one or some combination of six possible scenarios:

1. [**Primitive Forms**] The fundamental forces acting on all matter left this system of matter in its arrangement. We'll call these arrangements "Primitive Forms".
2. [**Replicating Forms**] The matter was absorbed from the environment arranging itself perfectly according to the information in its self-replicating molecule. We'll call these arrangements "Replicating Forms".
3. [**Life Forms**] The matter was absorbed from the environment arranging itself with variation according to the information in its DNA molecule. We'll call these arrangements "Life Forms".
4. [**Accidents**] Some being made this arrangement through an accident. We'll call these arrangements "Accidents".
5. [**Copies**] Some being made this arrangement to reflect a similar arrangement for the purpose of replicating the function of that similar arrangement. We'll call these arrangements "Copies".
6. [**Inventions**] Some being made this arrangement through invention. We'll call these arrangements "Inventions".

The first scenario describes how functional order occurs naturally through the effects of the fundamental forces acting consistently on all matter throughout time. The second scenario describes how functional order is grown from a self-replicating molecule. The third scenario describes how functional order is grown from a DNA molecule. The fourth scenario describes how functional order is created through accidental manipulation of matter by a life form. The fifth

Page 28

scenario describes how functional order (an arrangement) is passed on to other arrangements of matter (spread through influential order) through the manipulation of matter by a life form. The sixth scenario describes how functional order is created from reason through the manipulation of matter by a life form.

    The enlightening deduction here is that copies and inventions were created by reason, reason and accidents were created by life forms, life forms were created by replicating forms, replicating forms were created by primitive forms, and primitive forms were created through natural processes (the fundamental forces acting on all matter), so inventions and all other arrangements were created by these natural processes. These scenarios are all the possible reasons for any arrangement's original formation. Each of the numbered types of arrangements of matter represents another level of efficiency order has found throughout time for creating order that is more resistant to disorder. So matter continually finds higher order arrangements that are more resistant to disorder. This is the only reason any arrangement of matter exists. Everything we know of exists as an arrangement of matter, therefore, everything we know of is the result of matter finding higher order. The exceptions to this are the basic building blocks of matter, space, time, and the fundamental forces that act upon them. These exceptions created order, and order created everything else through them.

    Order is winning over disorder. What we do in every day life is find better ways to defeat disorder. We build orderly houses that better withstand the disorder in nature. Heat, cold, earthquakes, and hurricanes are all examples of material and energy that are applied to our orderly systems in a disorderly way. They sometimes break down our weaker order. Disorder, lower level order, is always chipping away at our higher order so we find ways of making more thermally efficient, structurally sound buildings that are resistant to disorder for a longer period of time.

The Meaning Of Life

The human being is the most evolved of the known species in our universe, largely due to its capacity to reason. We are the vessels of the highest known order and so we are constantly reshaping the matter of the Earth giving it more order relative to our desires.

Because we, as humans, have about 99.9% of our DNA in common, we start our lives very much similar. Our experiences, while sometimes vastly different, impart a very similar set of rules that guide us. These great similarities enable us to experience an object and often come away with a very similar impression of the order of that object. This similarity in relative approximations of an objects order enables us to often accurately decide what old order should be destroyed to create objectively higher order in its place.

However, since these decisions often have to be made by the few empowered with the authority to decide, the objective order of an object is often largely eschewed in favor of the level of relative order of that object in the minds of those few in charge of the space that object inhabits.

Fortunately, as our technology creates more realistic methods of experiencing an object, that object's order can better live on as digital information while the original object has long since perished. One of the ways we enable the creation of higher order is by finding smaller, cheaper, and longer lasting methods of preserving order we find functional.

Day after day more material is dug out of the primitive order of the Earth, reshaped, separated into more functional bunches, and given higher order for the benefit of mankind. We have taken over the far reaches of our planet and sent out precursors to our eventual exploration of our solar system and beyond. Order is winning by spreading through us, all the while finding more efficient ways of increasing itself by increasing its resistance to disorder.

Chapter 6. The Human Vessel Of Order

Now we get to the answer of the age-old question: What is the meaning of life? The meaning, or purpose, of life is to create higher order. I'm not saying we should or should not slave away in pursuit of higher order. I'm saying we exist because order is taking over disorder. It is doing so through our every action. It's not what we *should* do, it's what we *do* do. We have no choice but to try to create higher order. It is how we came into being, and even now it guides our every action. Order existed before us and we have come into being and evolved through order evolving. We are the first forms of order capable of understanding our order to the extent that we understand ourselves as vessels of order, both genetically inscribed in our DNA and learned through our experience.

With enough matter and energy being acted upon by the four fundamental forces for enough time, functionally ordered matter will find influential order through replication in a molecule with a blueprint for its own construction. It is thought that this order likely first took hold in the form of nucleic acids that have the property of replication but not variation. Once enough of this replicating matter exists, the disorder in its environment will introduce a bit of order that impairs that replicating matter's ability to replicate perfectly. This impairment proves itself functional as the variation necessary for evolution. This new form of matter, capable of replicating itself with variation, introduces the property of evolution that is present in life through DNA.

Evolution is the property of creating generational improvement. This property leads naturally to species that are continually better able to combat and survive the disorder of their environment. This leads naturally to the ability to reason whereby we anticipate and preempt disorder. In time this leads to life forms with the reasoning capacity of a human.

The Meaning Of Life

As a strand of DNA, in the form of a human genome, selectively collects and arranges atoms from its environment, it forms a child with sensors (eyes, ears, tongue, nose, extremities) connected to a central location where neural matter can form. This growth of neurons forms a network of connections that record the stimulus received by the child's sensors. As certain patterns of stimulus are received more often, the neural web formed from those patterns becomes strengthened. These patterns are rules, and the strength of those rules determines the level of trust the child puts into those rules.

Through this process we keep an imperfect record of every important thing we have experienced. We are always anticipating what we expect to experience so we can record how our actual experience differs from what we expected. This enables us to focus solely on what part of our experience surprised us so we can spend our limited neural matter recording what is needed to better anticipate a future experience and preempt disorder.

Once we've lived long enough to feel pain, even just the pain of hunger in the womb, we've found a need to satisfy, a reason to anticipate pain and avoid it. We're learning how to gain order by avoiding disorder.

Pain is the mind's recognition of disorder in a system that is important to it. For example, pain within our muscles as we exercise is from a muscle fiber's tearing from strain, which creates disorder within that tissue. This sensation of pain gives our conscious stream of thought a reasonable expectation of how much further we can go before we need to rest, allowing the regeneration (regaining the order) of our muscle fibers to the point that they can better meet our needs.

In growing back more muscle than we had before exercising, our body is responding to the fact that it may again be called upon to do this much work. We only add a bit more muscle mass to the places where it's likely to be used so as not to create too much body mass, which takes energy to sustain.

The Meaning Of Life

We have higher functional order when we can get the same result for less energy, which is the same as getting more for the same amount of energy. In this case we want to be able to exercise longer or harder, meaning we want to more easily do the same amount of work so we have energy (and muscle mass) to spare, to do more work.

    Emotional pain is no exception. It too is our mind's recognition of disorder in a system that is important to it. For example, the pain of losing a loved one is the recognition of disorder now present within our systems of family, personal history, support, and many others. At some point one may assume a system of the need to be a good son, daughter, grandson, or granddaughter. When our parent or grandparent dies we may wish we had called them more recently, to bring that system of order into better balance while we had the opportunity.

    If this emotional pain is difficult to bear, we develop new logic in our brains to better prepare ourselves for a similar loss. Maybe we remind ourselves more often to call the grandparents we have that are still with us. Perhaps we find a better way to understand and accept death. Either way, like gaining more muscle mass, we always try to and usually do leave behind a system better equipped to handle the disorder of a similar situation.

    We all go about the world trying to increase order. Not just by bringing temporary order to a system that grows more disorderly in between checkups, but by finding longer lasting order that's easier to achieve within those systems so that we can have more time and energy to take on higher systems of order or finding higher order in the ones we have.

    We maintain systems of order, arrangements of matter, within our mind, body, and environment in such a way as to make it easier to create higher order. Some people find higher order in their environment by putting everything neatly in its place so that it's easier to find when they need it. Others find it more useful to leave items out, in a messy fashion, where they're likely to be used, to save time setting up and putting down a project. Which method is of a higher functional order relates to the project and the preferences of the person working on that project.

Within the human mind lies a myriad of competing systems of order. Some come about from a need to fulfill the physical requirements that sustain life, which in turn sustain our ability to evolve higher order within our minds. Our food-energy, oxygen, and water systems are vital systems that have to be kept in acceptable balance for the mechanics of our body, the host of our order, to sustain itself.

Other systems we discover and adopt because we recognize, through experience and reason, their capacity to increase our ability to create higher order. Sometimes we adopt a new system through reason, accepting the cumbersome tasks required to maintain that system for the beneficial order it provides us. Other systems are programmed into our DNA so that we can satisfy those systems before we've gained a mature reasoning ability.

Which systems we choose to focus our attention on improving, depends on which systems have reached a low enough level of order, our estimation of those systems' importance, and the convenience of available tasks we can employ to bring about sufficient order within those systems. Through this mode of operation we exist as fleshy experiments through which order is constantly and efficiently increasing its ability to resist disorder by finding higher order.

We may now be the carriers of the highest levels of order, but we are just one incarnation of order in its drive to overtake disorder. We're not the first, and we won't be the last form matter will assume in its relentless pursuit of higher order.

As the highest functioning carriers of order, there are still many ways in which we fall short of our potential. Because, in most cases, we are the being most capable and willing to provide ourselves with the order we need, we live as somewhat selfish beings. As selfish beings, there are times that we have to choose between serving the greater good, and serving ourselves. We make these decisions by weighing the likely effect our actions will have on all of the systems of our mind that could be affected. The systems outside of our minds, such as our bodies and our environment, are only taken into account through our mind's impressions of them.

If we're of a mature form of order, then our experience with other people has left us with a desire to increase, or at least preserve, the order of other people. As beings of a mature order, we are capable of creating higher order for ourselves by acting to create higher order in others.

This system we adopt of the need to preserve or improve other people's order is an important system that I'll refer to as the system of society. If our system of society is immature, we don't care about other people's disorder, so we're capable of creating disorder for others to increase our personal order. Doing so increases our personal order at the expense of the objective order of the universe. When we fail to care about the order of another person we have assigned a low level of relative order to them. Often this comes from the misunderstanding that our own personal order is more important than the objective order of the universe. As we mature, we gain an understanding of the supreme importance of the progression of objective order.

Because a system can be assigned a level of relative order very different from that system's level of objective order, many of us find it hard to believe that we all go about the world trying to create higher order. For example, without an understanding of how much pain person A has caused person B, we have a hard time understanding how, in the heat of passion, person B could think they're creating higher order by murdering person A.

The truism that "time heals all wounds" has wisdom that is proven out here. Immediately after person A causes person B pain by inflicting a wound on person B's systems of

The Meaning Of Life

order, the new disorder is most painful. As person B continues to live, he/she builds order within his/her mind or body to overcome that disorder.

In most cases a wise person B will "turn the other cheek," knowing that they will rebuild the disorder of their wounds over time. By reacting in the heat of passion we spread the disorder of our wound to other systems by sacrificing order in our many other systems for temporary relief in our wounded systems. We may also sacrifice the level of the relative order we impart on other beings that is their impressions of us. This can create an environment where it becomes increasingly difficult to get the order we need from others.

We are all acting to create higher order. Some of us do so in a wise way that sets ourselves up for easily finding future order in other systems. Others are not yet functional enough to delay the gratification of higher order that we all want to receive. This is not to say however that turning the other cheek is always the wise move. In some situations, time is of the essence, and the moment after a wound is received must be seized. This can often be done in a wise manner by ignoring all of the disorder of the recent wound except for the knowledge that this being or system is capable of inflicting such disorder, when reacting, so as to not react rashly.

Most of our decisions do not pose a choice between bringing greater order to our selfish systems of order or to our system of society. Most choices we make are between tasks we can perform to bring about the highest possible order to our system of subsystems, which leaves behind a higher ordered human race.

We have adopted our systems of society as a means to help objective order persevere over relative order. We have done so because order finds more orderly ways of increasing itself.

Since the Big Bang, all the matter in our universe became more orderly through the constant application of unchanging forces until it eventually found evolution through the DNA molecule. With enough time, order, in the form of DNA, found an arrangement of itself within its environment that could create a new type of environment where evolution could happen on a much faster scale; the brain.

In doing so, order didn't have to find a new molecule capable of evolution. It took the easier route of finding a new environment, more hospitable to DNA's evolution, and a body to do the necessary tasks of manipulating and consuming its surrounding environment to sustain the orderly environment of the brain. The biological mechanisms through which memes are formed within the brain are understood to some extent today. For example, life has evolved what we call the CREB (cAMP response element-binding) protein, which provides the mechanism by which our perception directly affects genetic expression and new memory formation at the cellular level.

We can perceive light as photons through an array of photon-sensitive receptors in our eyes. Each receptor can receive light in either the red, green, or blue wavelength. These receptors send a signal corresponding to the number of photons they've received of their wavelength to our brains, where our CREB proteins can receive that signal and switch on or off our creation of certain genes, strands of DNA, that form neural matter. This is a simplified way to describe how something we see becomes a memory stored in our brain. We process these memories, extracting useful information in the form of memes that are subject to the same laws of evolution that have determined our biological structure. Our neural memes, however, exist in an environment where their arrangements can evolve much faster than their biological counterparts.

This biological process of storing image memories is remarkably similar to how our digital cameras work today. In digital cameras there is an array of photon-sensitive sensors for red, green, and blue light that can send a signal to the camera's processor telling it how many photons of the corresponding

The Meaning Of Life

wavelength each sensor has received. The processor uses an arrangement of transistors to logically determine whether to write a 0 or 1 (on or off) to a location in memory that stores the image.

The other senses we possess have also been synthesized in a similar fashion through our technology. It may appear that this technology is evolving at a faster rate than we are, but actually we are evolving through our technology. Our technology serves us and is the fastest way for us to adopt and evolve new functional systems of order.

Our biological material is more complex and therefore is a more difficult medium in which to create functional order by hand, however we are in the early stages of a biotechnological revolution where we will better master that complexity. As we do so, our bodies will become our technology as we further our resistance to the disorder of age, wounds, and disease.

Chapter 7: Level Of Order As Importance

As human beings we are all carriers of unique systems of order. Because of our unique systems of order, we have unique pieces of order we're looking for, unique desires to be fulfilled, to further our order. Because of our unique desires, any system can have a unique level of relative order for every being that experiences it, but just one level of objective order. As beings with limited capacity to understand the level of objective order of things, we approximate those things' order and only break their current order when we think we can create higher order by doing so.

One person can see a new work of art, by an unknown artist, as a thing of beauty, while another sees it as a piece of trash. That work of art will have a high level of relative order for the former person, and a low level of relative order for the latter. However, there still exists the level of objective order of that piece of art, which is the accumulation of relative order imparted on everyone who has experienced that work of art.

The level of relative order for some entity can be more easily understood to be a person's approximation of how important that entity is. An entity's objective influential order changes throughout time as different people are exposed to it, and as the memory of that work in the minds of those who have seen it fades away, or is remembered and perhaps reflected in their own creations.

This illustrates the impossible difficulty of perfectly determining the level of order, or importance, of something. It would require perfect knowledge of every brain containing relative order from that something, and the context those brains exist within. Understanding the level of objective order of anything perfectly requires a perfect understanding of everything in our universe.

Because humans are the creators of forms with the highest known levels of order, what is important to us is

The Meaning Of Life

usually important to the progression of order, but not always. For example, when an influential person such as Adolf Hitler acts in a way that temporarily furthers the order of himself, or his country, by destroying the order of many other people, his actions may have the effect of slowing the progression of order.

A system's level of order is only equal to its importance if we understand the progression of order to be the only thing that is important. Order existed before life and will exist after the human race, as we know it, is gone. What is important to us humans will not forever be what is important to the progression of order. Also, because many of us are more concerned with our own progression than we are the progression of the human race, and because we can only approximate the level of objective order of things, we are not always acting to further the highest currently known levels of order. However, the vast majority of actions we take to further our own order, end up improving the order of the human race. Therefore, when trying to understand order, it can be intuitive to associate a level of order with importance.

With this in mind we can get a clearer picture of what order is. If something is important to me, I have given it a high level of relative order. If something is important to the human race it has a high level of objective order. Because all order starts as functional order and can then attain influential order, a system of order can have high functional order, high functional importance, without yet having high influential order.

For a person to receive a piece of functional order and give it a high level of relative order, that person must be at a high enough level of functional order to understand it (to put it in a highly functional context), but not such a high level that the new piece of order is redundant. We are all looking for a piece of order that we can comprehend, and learn from, to increase our order. If we receive a piece of order that is too high or too low in functional order we will give it a low

The Meaning Of Life

relative order because it is not important to us if we can't use it.

I raised the example of Hitler because it illuminates a couple of aspects to a system's level of order that are difficult to grasp; Did Hitler have a high level of order? If all of our actions are meant to create higher order, then how do you explain Hitler?

To approximate a person's level of order you have to take the totality of their actions, (order created minus order destroyed), and weigh the resulting order they left behind. Hitler's example is one of history's most clear-cut cases of a person whose destruction of order likely outweighed his creation.

This is a case where the level of order does not equal importance. Hitler was undeniably an important figure. He left his painful mark on humanity. However, he most likely did not have a high level of order. Once again, a system's level of order is only equal to importance if we understand the progression of order to be the only thing that is important.

Hitler's actions, like all of ours, were meant to create higher order. He misunderstood the importance of his own race, and of race in general, in determining a person's level of order. This led him and many other brutal dictators in our history to commit genocide.

He had a high level of relative order for members of the Nazi party and their supporters during the age of the Third Reich. He had a very low level of relative order to almost everyone else during that time. In the end he did damage to his supporters, detractors, and victims.

Of course, accurately approximating a person's level of order is much more difficult than this. One would have to consider the technological advances created during World War II and Hitler's role in facilitating them. Most of the credit for this order would go to the scientists and engineers themselves though, as well as other leaders for facilitating their country's technological advances. Hitler may have created so much

The Meaning Of Life

disorder that much order could rise out of the ashes, but the hard work and ingenuity necessary to create that new order came almost entirely from other people.

There is also the historical lesson of Hitler to consider, which is a very useful example to describe how a people humiliated and thrust into an economic depression can perpetrate terrible disorder out of desperation. This lesson will live on in perpetuity to remind us of what evil we are capable of and how to avoid it.

This lesson is of a high level of order. It will resist disorder for its capacity to teach us all how to avoid great disorder in the future.

Chapter 8. Tasks

A task is order that serves as a recipe for bringing about greater order in a system. Whenever disorder within one of our systems is sufficient to demand our attention, our stream of thought analyzes our known applicable tasks like a plumber looking for the right wrench. We then schedule the most orderly time slot for that task to be performed.

In this way we are very much like the machines we create. Computer operating systems schedule tasks with an associated priority to be carried out for the creation of higher order within the user. We imagine ways in which we want to manipulate information and use the willing man-made mind of our computer to do the tasks necessary to bring about the order we imagined.

Through our creation of a machine to do some of our tasks we are able to make a mind much simpler than our own with minimal needs that has more functional order than ourselves at its intended tasks. Today, our personal computers are still a lower order system than us humans for the tasks of finding sustenance from the environment. They are not yet able to procreate in an environment without human beings. They are still our technology. Their form of life is like a successful meme, requiring the willing host of mankind to recreate and sustain them. Man may use computers to build other computers, but should man cease to exist, computers, in their present form, would not be capable of harvesting the necessary materials from their environment to persist for long without us.

Computers will likely reach this capacity, but now they act in a purely supportive role, adding to the order of their human operators. Humans make the important decisions, but their decisions are often aided and improved upon by the powers available through computing.

The Meaning Of Life

With all of the order necessary to self-sustain in our environment, the order of the human being is still unmatched by any other vessel of order. We have evolved as vessels of order most capable of furthering our order against the disorder of our environment. Through understanding our systems of thought and the tasks used to sustain and grow order within those systems we find the fastest ways to higher levels of order.

By analyzing the frequency we're called upon to do a task, we can focus more time and energy on improving our most commonly executed tasks. This is why the wisest person in the world will not be able to tell you when the next train will leave for your desired destination. He/She would rarely be called upon for such trivial information. You'll find the highest functional order for that task in the average train station attendant.

This alone should explain the fallibility of some of our celebrities, CEOs, and politicians. To reach the top in our chosen profession we often spend a great deal of our time improving our tasks that better order the systems required by that profession to the detriment of many other systems of order we've adopted, such as being a good parent, spouse, friend, or member of society.

Doing this is simply adopting a specialty. When we choose to adopt a specialty we've discovered a system, or group of systems, we're particularly functional at achieving order within. Perhaps we have discovered a highly functional task that we can employ to provide others with some order they desire using little time and energy. Perhaps we have found an interest in some task sufficient to make the work necessary to master that task worthwhile. Perhaps we've better ordered our muscle to do a strenuous task with relative ease. When we choose a specialty to adopt we increase our reliance on our community, while increasing our value to that community.

This is where influential order comes into play. We don't have to be capable of finding order within all of our

The Meaning Of Life

needy systems on our own as long as we have sufficient access to someone who can, and are functional enough ourselves at some other task as to have something to offer in exchange. Access is a property of our order. If we have encountered someone and approximated that person to have a high level of functional order at a certain task, then they have imparted relative order onto us that we can use to have that task executed as long as we maintain access. The degree of order we impart upon them helps to determine the level of access we have to their order. Just having some money provides us with access to most goods and services. Building close relationships can provide us with access to many things unattainable with money.

When we participate in an exchange, we've recognized a win-win scenario. The win-win scenario is the realization that one man's trash is another man's treasure, which is the natural result of different carriers of order needing different types of order at any given time. These scenarios are abundantly present around us, and the most successful people are ever cognizant of them.

Through the realization of a win-win scenario one life form can give another something to receive something of greater value, while the other life form does the same. The rate of exchange of currency is a dominant factor in the success of an economy. The more people are spending, the more win-win scenarios are taking place, and hence the more order is being created. These exchanges improve the order within both life forms and are the primary mechanism by which we exist as social and commercial beings.

Some practical advice can be derived from this for those seeking love. Don't spend your time looking longer or harder for the right person. Instead spend your time building higher order in yourself to give you the confidence that when you find a potential partner you'll be able to approach him/her offering a win-win rather than asking for a favor.

Chapter 9: Finding Wisdom

Now that we have a common understanding of order, its parts; functional order and influential order, ourselves as vessels of order, and tasks as functional order we execute and improve upon to bring better order to our systems, we can focus on how to find higher levels of order.

There are three fundamental capacities for finding higher levels of order that I will discuss in this book: knowledge, wisdom, and creativity. The processes of obtaining the last two warrant their own chapter, whereas the process of obtaining knowledge can be described in a sentence. To obtain knowledge, we simply need to seek out information and memorize it.

To accrue wisdom is to find a better context for the knowledge we have. As we accrue knowledge we put it in the best context we can on the fly. We usually have some new piece of information we want to remember presented to us soon after the previous piece, so we expend limited effort trying to understand the information we receive.

When we spend time after a learning session recalling our new knowledge and interpreting it into a more functional understanding, we are finding a better context for that knowledge and gaining wisdom.

In a practical sense we can find wisdom through a simple exercise. Take any concept you come across and try to describe it in the context of order vs. disorder.

Recently I was on a subway car in Paris and I noticed the word "envy" scratched into the window next to me. I began to wonder how the concept of envy relates to the progression of order over disorder. Often, in my past, when the word envy has been invoked it has been used in a cautionary way, as some demon to be exorcised from our minds. It's treated as something that can only do harm, but I

The Meaning Of Life

wondered how could it persist within our nature if all it served was disorder.

I then thought perhaps envy is the recognition of some higher order within another being that creates a desire within us to remove that inequality. There are two ways in which we can equalize our order with that of the being we envy. We can learn from the other being and find that order for ourselves (i.e. "Where did you get that wonderful coat?"). Or we can destroy that order (i.e. "Sorry I spilled wine on your wonderful coat.").

Given that it takes less time, energy, and imagination to destroy order than it does to create it, perhaps it was deemed wise to render guilt upon those who envy a certain order so as to preserve it from destruction. Here, with our better understanding of order, we can see the purpose of envy.

When we simply admire something, perhaps we'll want that order for ourselves, or perhaps we are satisfied with it simply in our memory. When we envy something we want it. We need not feel any shame for our envy though, for envy is a mechanism by which high functional order gains influential order. We want something and if we understand the productive method of channeling that envy, we'll use it to our advantage and buy a similar coat, spreading the influence of that design.

This example illustrates how there is an exception to almost every rule, if not indeed every rule. Rules exist as concise understandable pieces of order. For a being possessing a relatively low functional order of the system to which a rule applies, the rule exists to be obeyed. Most of the time, when children obey the rules they are increasing order within themselves and the systems to which the rule applies. Accordingly, when children break a rule they are most likely creating disorder within the systems to which the rule applies, though they're likely increasing their functional order (knowledge and wisdom) of the systems and the rule.

People with high functional order of the systems to which a rule applies are more likely to understand the reasons the rule exists therefore understanding its exceptions. When they break a rule according to its exceptions they will likely create higher order within the systems the rule applies to, and within themselves. They will likely create higher order than had they followed the rule.

For a rule to be followed it must be remembered and to be remembered it must be in a concise form. In a concise form it must exist without its long list of possible exceptions. To quickly gain higher functional order about any system we should always question why the rules that apply to that system exist. If there's no one who can answer these questions we can find ways of breaking the rules without doing severe damage to the system. By observing the resulting effects we better understand the rules and the system to which they apply.

A rule is like a gene within a strand of DNA. Both exist as our most concise known piece of order to guide order's creation of higher order. A gene exists as a series of letters within a four-character alphabet (C, G, A, and T) that describes an orderly path for matter to take on more matter so as to form the function of a nose or an eye depending on the gene. A written rule exists as a series of letters, within the less limited character set of the alphabet of the language in which it was written, that describe an orderly path for people to take.

The exceptions to the rule of a gene that create higher order are found by that gene's imperfect replication, a crude almost brute-force method of trial and error. As Darwin has shown, a genetic exception that improves its host's ability to survive, will itself survive.

The exceptions to a written rule are found through the more orderly application of trial and error by humans. Humans, through our powers of observation, reason, assumption, and manipulation are able to make one error and deduce the likely error within similar possible exceptions without trying them. These powers, most functional in humans, evolved naturally over millions of years and we are still evolving new powers.

The Meaning Of Life

For example, through our technology, we've recently evolved the power of computer simulation. Using this power we build mathematical models of reality and test exceptions to the rules in our designs. Through this power we can find the flaws in our designs faster, with more accuracy, and less expense.

This power, like all powers before it, is a unique capability created through the natural evolution of order, but it is more than the average generational increase in order. A new power is like a new species. A system of order evolves incrementally, requiring slightly less time and energy to achieve the same result, until it reaches a threshold functionality that enables the user of that system to get a new, more functional, result.

Before we built these speedy, accurate computer models, we sketched these models on paper using similar, sometimes exactly the same, mathematical rules. However, the time it took to do certain calculations by hand prohibited us from accurately predicting the effects of hurricane force winds on a structure. So we built smaller scale physical models in a wind tunnel and added unnecessary weight to the final structure to compensate for our imprecision.

Before these rules were defined in the language of mathematics, our observations and assumptions did a pretty good job of approximating answers to many of these same questions we now ask of our supercomputers, yet we had to accept an even higher risk of failure.

We create generational advances in the order of our systems by modifying their order so that we can receive the same result from that system using less of our time or energy. Eventually, a new generation will be able to solve problems that the previous generation found prohibitively expensive or in some other way, insufficient. Some of these advances prove so functional that they produce a revolution, yet all of these world-altering moments come from incremental progress on an existing base of order.

The Meaning Of Life

If we think of wisdom as being of a mind with a high level of functional order, then when we seek wisdom we're often seeking a way to get the same things we've been getting with less effort. Of course we want something new to surprise us as we seek wisdom, we want that "Ah-Hah!" moment where something new makes sense, and yet we find those moments at times when we're better ordering what we already know. We can receive new functional order, new information, the missing piece to the puzzle, but until we've assembled the puzzle sufficient to realize a piece is missing, we won't know to look for the missing piece.

This is because a higher form of order can only grow on a base of lower order large enough to support it. Most of the work it takes to form a higher pinnacle of thought comes from expansion on the base of that pinnacle. Once we realize something new about a system we begin applying that knowledge more loosely to similar systems.

Therefore, if we wish to gain greater functional order (more wisdom) we can recognize when we've learned something new and be more thoughtful of how that lesson applies to similar systems. As that lesson spreads throughout our various systems of thought it creates room for an exception to that thought that we might find functional enough to, again, spread to similar systems.

In the same way that we create more structurally sound materials to resist the disorder of the environment, we create more logically sound arguments through success and defeat in a battle of ideas. Through communication and argument we better hone our system of rules to the situations to which they apply. This enables us to recognize which rules are of a lower order, needing to be better defined to leave behind a higher ordered system of rules.

We can all improve upon our own understanding of things by removing any emotional attachment we have to our own set of rules. This emotional attachment often keeps us from accepting a better argument that comes from someone

The Meaning Of Life

else. When we are more concerned with being the one recognized for having the correct answer than we are concerned with finding the best answer, we stunt our own intellectual growth, and frustrate those whom we could learn from.

Chapter 10: Finding Creativity

Creativity is the ability to functionally apply lessons learned from one system to a very different system. If you want functional order that is hard to find, you should ask a knowledgeable person. If you want functional order that is accurate and insightful, you should ask a wise person. If you want functional order that has not yet been discovered, you should ask a creative person.

      Creativity is a combination of wisdom and risk. To be creative is to accept a high potential for failure to find higher levels of order through correlating dissimilar systems. When we have knowledge that is very functional in its context within system A, we have some wisdom about system A. When we recognize similarities between our wisdom of system A and our wisdom of a different system B, we can use our creativity to correlate our wisdom of systems A and B.

      To functionally apply the wisdom of system A to system B we need to find aspects the two systems have in common. We can then analyze the wisdom we have for system A that applies to those shared aspects of A and B, applying only that wisdom to system B.

      This will produce creative insights into how we can modify system B to give it the desired functional order we could extract from system A. Through these modifications, we create a system C that hopefully has the desired aspects of A and B.

      Imagine when looking out a window we see a uniquely beautiful bird posing for us and we wish to preserve that memory. We can recreate the colors and form of the bird in the window (system A) through the corresponding colors of paint and strokes of our paintbrush onto our canvass (system B). We end up with a painting (system C) that combines the beauty of the bird with the stationary permanence of the art supplies.

The Meaning Of Life

    Of course our creation may fall short of the beauty we saw in the bird. That is part of the risk we accept when we are creative, but we learn from our creative process so the next time we try a similar project our brush strokes improve, our colors are more accurate to our purpose, and the order we create is more functional.

    The wisdom here was a simple understanding that the bird is unusually attractive and that we can replicate color and form using art supplies. We understood that the functional aspect of the bird was the image of color and form it left in our mind. Using our system of art supplies along with our wisdom of how to apply them to replicate, and perhaps exaggerate, the interesting aspects of the bird, we use our creativity to make new functional order.

    When we use our creative ability to make functional order, be it a work of art or a scientific theorem, we imagine some unknown form containing ideal aspects that are a combination of aspects we have experienced in various other forms of order. We then attempt to create a form with those aspects. The creation can surprise us, revealing itself in more functional ways than we had intended. This is one of the many deeply satisfying results of creative experimentation.

    We all have this capacity within us, but many, when faced with the choice of spending time and energy to experiment in an unfamiliar medium for uncertain results, or spending that time and energy doing a known task with a known reward, will opt for what is certain. Risk requires a mature perspective of failure.

    Failure causes some disorder, but the process of creation exposes us to high levels of order regardless of whether the final product is a failure or success. Failure causes some disorder, but so does the natural passing of time. Every carrier of order will, one day, fall back to the primitive order of the Earth. For a living being, eventual disorder is inevitable, but the disorder of failure is temporary and often necessary for the creation of higher order. If we learn to embrace our failure then we disarm it of the disorder it can cause.

When we begin to understand our failures like Thomas Edison understood his unsuccessful attempts at a practical light bulb, we see them as a piece of our eventual success. This is a mature understanding of failure.

The boundary between creativity and wisdom is not neatly defined. When we put any functional order into a context we do so by correlating the functionality of similar functional order in a similar context. We assume some risk that this functional order is more functional in a different context.

To put functional order into a functional context within a system, to create wisdom, requires an understanding of the many aspects of that system. To correlate two very different systems, to be creative, requires an understanding of the many abstract aspects of those systems. So to foster creativity we should encourage the comparisons of apples to oranges.

To do so gives us a better understanding of what an apple is and what an orange is. We understand that they share the aspect of being a fruit, but have different aspects of color. Furthermore we understand that fruit is edible and sweet, so if someone wants an apple and we don't have one, we might improvise by offering them an orange because they share many similar aspects. When we improvise we are using our creativity to find a solution with what is easily available.

When we develop our creativity, we also develop an abstract wisdom, which is a higher form of wisdom. Wisdom of the abstract correlations of order gives us a deeper understanding of any system of order and its level of functional order. We can use our abstract wisdom to correlate all systems of matter providing us with more functional methods of putting our knowledge into a more functional context. Which is to say, our creativity exposes us to greater wisdom.

When seeking wisdom or creativity, it is important to understand the interdependence of knowledge, wisdom, and creativity. It takes knowledge to form wisdom, and it takes

The Meaning Of Life

wisdom to form creativity. You don't need wisdom or creativity to gain knowledge. You only need to seek it out. Careful consideration of knowledge will give you wisdom. Abstract correlations between wisdom will give you creativity, which will give you better wisdom.

This does not mean that creative people have more wisdom than wise people. You can be creative with the smallest amount of wisdom. Creative experimentation is, however, a functional path towards finding wisdom.

When someone says you're comparing apples to oranges, they're suggesting that it is unwise of you to compare the two systems, and they may be right. Whether it is wise or not depends on the purpose of the comparison.

When we want to find wisdom (a functional context) from some knowledge of a system, it is best to use apples to apples comparisons. This minimizes the disorderly effects of unknown aspects of the systems we're comparing, helping us to find the best context for some knowledge.

When we want to find a creative solution from some knowledge (that is hopefully in a wise context), we compare apples to oranges because all apples to apples comparisons have been exhausted, or none can be found. When we want a creative solution, the obvious solutions are insufficient. For knowledge to be useful for creative correlation it must first become wisdom through careful consideration. When we observe the level of functional order within our creative solution, we gain new wisdom from the accuracy of our abstract correlations.

Chapter 11: Education

Education is a long complicated process by which one generation tries to best pass on its order to the next generation. It's such an important process and there are so many opinions about this process that our system is an amalgam of competing principles and ideas that do a pretty good job at educating our children.

There is no perfect method of education, just as there is no perfect method of governance. We have to continually try to improve on a process with just adequate metrics of our success. If our systems of measuring progress become too rigid, we harm the process of education by unifying around a single education system so our measurements can become closer to an apples to apples comparison. This stifles ingenuity that creates progress within our systems of education. On the other hand, if we become too loose in our accountability, we impede our capacity to appropriate funds on a basis of performance, which also stifles ingenuity by not rewarding good ideas and hard work.

This balance between accountability and flexibility is where much of the debate about education is centered around today. While this is an important debate, it's not one that an understanding of order will bring immediate insight into. It is a debate to find the most practical level of accountability to work most effectively for the current system. There is, however, another aspect of our education system that can be greatly improved through an understanding of order.

Top down vs. bottom up: Any time we are trying to understand a large system of many subsystems we employ certain methodologies that are of either a top down or a bottom up variety. To analyze a system using a top down methodology we first understand an overview of the system with only brief attention to its many subsystems. We then analyze each subsystem in greater detail. A bottom up

The Meaning Of Life

approach involves the study of each subsystem in detail as we slowly build towards an understanding of the entire system.

In the United States, our current public education system is based almost entirely on a bottom up approach. We divide the day into classes that teach the various subjects we deem to be critical. In these classes we, to a large degree, learn the many concepts of math, literature, art, science, etc. in a vacuum. Often a child will notice a similarity between concepts learned in different subjects and if we are fortunate he/she will see some logic behind their similarity. If we are unfortunate, he/she will disregard these important connections as coincidence.

A child could draw a tree that has lost its leaves in art class and notice how similar it looks to the circulatory system around a pair of lungs he/she is studying in biology. These same structures could inform a discussion in history class about the success of certain military supply routes. These important similarities are not a result of mere coincidence. They are instructive of an optimal path for the transport of material (water, blood, military equipment) through a branching network hierarchy. Examples of the abstract correlations between the primary fields of study reveal creative approaches to problem solving that can be a catalyst for finding order of the highest levels.

The primary reason we have failed to teach from both a top down and a bottom up approach is that we have failed to agree on a paradigm through which we can see all things as connected. Another reason is that not all of us yet understand everything as being connected, or at least disregard the utility of such a view. Any challenge as large and complex to solve as how to pass on our order to our children requires all methods of attack we have at our disposal.

When a person wants to be elected president, he/she must mobilize a movement of people to support his/her candidacy (bottom up), while convincing the current instruments of power that he/she is acceptable (top down).

The Meaning Of Life

This hybrid approach provides many self-complimenting factors. For example, when powerful politicians see a wave of grass-roots support, they can gain some of that support by joining that movement. Likewise, when respected people in the upper echelons of power lend their support to a candidate, it garners attention from supporters of those well-respected people.

Another example of this hybrid approach is the successful execution of a war. To win a war responsibly is no longer seen simply as defeating an enemy. To win a war requires capturing territory and then the hearts and minds of the masses through a powerful but fair use of force and resources. Success requires attention and resources flowing to local authorities with respect paid to the average citizen combined with high level diplomacy and aid. It is much easier for the bottom and top to meet in the middle than it is for either side to force the other into submission.

This concept is just as applicable with our system of education because of the nature of such processes with immense scale, competing principles, limited accountability, and regional variation. Such a complex process requires multiple complimentary methods of execution to cover the gaps in any single approach.

If we can agree that a hybrid approach is preferable we must unite around a general understanding of how these subjects are interconnected to form a top down approach. An understanding of order vs. disorder can provide this connection. To illustrate this I'll provide a brief overview of a few of our most common subjects as they relate to order vs. disorder:

Science: The effort to discover or to better understand the order, natural and theoretical, of the universe and its many subsystems.

Math: A subset of science that relates to theoretical analysis of quantity, structure, space, and change.

Art: Order created for the effect it has on human emotion.

History: Order in the form of knowledge that serves as a record of what has happened in the past.

Language: Order as arrangements of letters that empower mankind with the capacity to record and communicate high levels of order.

Physics: A subset of science that studies how matter behaves naturally in the universe.

Chemistry: A subset of physics confined to the study of the properties of matter in forms with low levels of order; elemental matter, and molecules.

Biology: A subset of physics confined to the study of the properties of matter in forms with the high levels of order that constitute life.

Psychology: A subset of science confined to the study of the effects of matter in forms with the highest levels of order, neural matter.

Physical Education: The study of maintaining a high level of order within our bodies through active participation in exercise.

     These definitions are not perfect, nor are they complete. They are merely a beginning. For example, when defining science, math, and physics, I used the terms natural and theoretical to create a distinction between order of the universe and order conceived within the mind. Our minds are part of the universe and therefore anything theoretical is also natural.

The Meaning Of Life

The distinction can still be useful, but its imperfection should also be understood. Our teachings should not shy away from the imperfections in our wisdom, be it a definition or a theorem, for those imperfections are often where higher-level order is found.

One reason why an understanding of order vs. disorder is a useful approach to a top down education is that it provides a first principle through which all things can be derived. Once we have a cursory knowledge of many things interrelated through a logical paradigm, when we learn something specific about one thing, we'll have a logical route to determine how to correlate that knowledge to similar systems. Our children won't build a huge knowledge base and only then begin to trust in their ability to make the high order conceptual connections between their pieces of knowledge. Every new lesson will have its logical place in relation to all other lessons.

Chapter 12: The Future Of Humanity

While the future is impossible to predict perfectly, through an understanding of order, there are many obstacles to the existence of the human race that can be predicted with great probability.

The first obstacle to the existence of the human race is to defy the very nature of order by resisting technology that changes what it means to be human. This will prove to be an insurmountable obstacle as it has been for every level of order before us in countless feeble attempts to pause the progression of order. However, given the fact that we are now beings of order capable of understanding order, and ourselves as vessels of order, we are uniquely capable of blazing a new path to higher order through acceptance and assimilation of this technology.

Just as large corporations are relentlessly resistant to change, so is the human species as a whole. We all look out to the world and see all kinds of changes we would like it to adopt while we fiercely resist the world's attempts to change us. We all live within our own bubbles of reality that we have spent years adapting to. Change in one system often causes disorder in that system and other systems that were built around the old system. Change is required for order's improvement, but it often takes work to build the order necessary to compensate for the change. When we are exposed to the disorder of change, we can experience pain, which is our mind's recognition of disorder in systems that are important to it. If the system that is changed is unimportant to us we won't feel pain from it. If the change works to our obvious advantage we recognize it as higher order and therefore not painful.

Large corporations are resistant to change because they have become large and successful by building their rules to most functionally operate in the current environment. Once that environment changes, so does the functionality of their system of rules. Likewise, whatever species is at the top of the food chain will resist being overcome by an emerging, more functional, species with life and limb, but will ultimately fail.

The one rule that persists over the resistance of the highest current form of order is the rule of order itself. Order is constantly finding higher order. This higher order will always come to dominate lower forms of order by its very nature. Keep in mind, we humans did not create order, order created us, and will surpass and dominate us no matter how hard we try to resist it.

So while large successful corporations are resistant to change, they only resist the change that they have the influence to restrict. To achieve and sustain success, businesses have to be constantly adapting to their changing environment. To do so requires creative leadership that can not only create a system of rules that are extremely functional for the current environment, but also periodically question the relevance of even the most basic of those rules.

Our technology has given us the power to simulate reality to better understand what actions we should take to most functionally bring about the change we want. We use this power for the construction of our buildings, roads, bridges, electronics, pharmaceuticals, and biomedical devices, to name just a few.

As we merge our biology with our technology we are speeding towards a post-human existence. Currently, some of us opt to remove specific genes from the genome that becomes our child to eliminate genetic disease, though this manipulation is currently limited and experimental. Soon our comfort level with genetic manipulation will reach a point that we deem it appropriate to use for functional enhancement not limited to disease prevention.

Disease ridden hearts are replaced by synthetic hearts that will continue to progress until the synthetic version lasts much longer than a human heart. Synthetic limbs will progress

The Meaning Of Life

until they surpass the capabilities of our best athletes. We will have to accept these athletes into our sports when the number of "pure" humans dwindles, and our concept of "purity" progresses to a form of bigotry.

We will isolate genes responsible for the development of ideal intelligence, strength, and beauty, giving parents the choice of raising an old-fashioned human, or a more successful post-human. Our human vessel of order will become, and already is to some extent, our technology.

While our fleshy existence gives us all many capabilities not currently available in our technology, our technology will soon surpass all of these benefits, and we will assimilate new devices around and into our flesh to stay at a competitive level of order. We will continue to do so until man and machine are indistinguishable, ever evolving away from the form we now recognize as man into whatever functional carrier of order is most competitive in its environment.

Through our technology we'll be able to greatly prolong our life as well as our youth. Eventually it will become possible to essentially pause our aging process at what we deem to be an ideal age. As we better understand the order of all of our cells and their systems we'll develop internal machines to monitor and repair our cells according to our wishes.

Our eyes will have the ability to zoom, see in spectrums outside the visible light range, and overlay other information we find useful. Our ears will hear a wider band of frequencies at our chosen sensitivity without degrading from age while filtering out noises loud enough to be destructive or just distracting. Our noses will be set to desired levels of sensitivity, from off to better than that of a dog.

The human body is the most powerful known vessel of order in existence. It is the housing of our brain. Its purpose is to sense and manipulate the universe outside of our brain according to its desires. Any way in which we can better sense

The Meaning Of Life

or manipulate our environment is a power we will seek for ourselves to better create order and avoid the disorder around us.

There is another near term obstacle to human existence that will soon demand from us great change and sacrifice. Global climate change threatens to alter the properties of the Earth that make it so uniquely suitable to life. Our environmental problems go beyond carbon emissions, though that is widely believed to be the most pressing challenge to a sustainable future.

Humankind, through its accelerating mastery of its environment, has found many forms of chemical energy to consume in its pursuit of higher order. We have done so with great disregard for some factors in the equations that describe the transformation of chemical energy into mechanical, or electrical energy.

At first, when we created the internal combustion engine, we marveled at its usefulness in the first automobile. This highly functional piece of order quickly became influential, sparking a consumer demand that fueled our creation of new methods to make cars cheap enough to produce in mass quantities. This made it even more influential. Soon it became a necessity for many of us to have a car so we could have a job and stay competitive. This advance in technology greatly increased the order of those able to afford it by exposing them to new people, places, and things, and therefore, new order. It became a powerful mechanism through which order could increase itself, so we have become increasingly dependent on it.

We so loved our new mobility that we paid little attention to the other product of our increased burning of fossil fuels. Before long, the knowledge spread that we shouldn't leave our cars on for very long in a closed garage due to the harmful fumes they emit. All we needed to do was to open the garage door so the fumes could be whisked away by our environment. Now we're beginning to see our environment more accurately like a big garage.

As we upset the natural balance that has sustained our Earth's chemical equilibrium, we are trapping more of the heat coming from the sun, killing off lower forms of life that we depend upon as a food supply, and depleting our fresh water resources. At the same time we're poisoning the air and water necessary to our survival. We are fast approaching a point of great danger for the survival of our species.

While our non-living environment is correctly seen as disorderly in comparison to the flesh of a living being, it is remarkably orderly in comparison to the environments of all currently known planets. Naturally, the Earth's environment is especially orderly for the tasks of supporting life on Earth. This is because Earthly life forms are orderly systems of rules that have proven functional in their environment, Earth.

Because of the relatively high order of the Earth's environment in comparison to other planets, it is imperative that we all recognize this common need of humanity as everyone's individual need to make the sacrifices necessary to bring our environment back into balance. Though we will one day be self-sufficient on other planets, giving the human race a firewall against catastrophe, the day in which global climate change threatens our existence appears to be much closer.

Through our evolving order we will be capable of finding clever ways to adapt ourselves to different environments on different planets, and even to adapt those other planets' environments to our needs. This is quickly becoming the most pressing obstacle for human order to continue its improvement on Earth, so we will refocus our work towards finding ways of affecting the Earth's environment, and adapting ourselves to its growing disorder. Doing so will give us a necessary knowledge base to build upon for terraforming other planets and sustaining ourselves on them.

The Meaning Of Life

As order progresses into the reaches of our solar system, the highest carriers of that order will increasingly look past their perceived barriers of expansion to find ways to reach other stars. Just as early explorers conquered the vast expanse of our oceans, and more recently the distance to the moon, order thrives from its capacity to consume more matter.

The form in which interstellar travel will prove possible is unknowable at our present stage of order, but from an understanding of current trends in technology, the idea is feasible in our future.

Currently we are in the information age where the technology of our computers and their Internet reins supreme. Over time, we are better mastering our ability to manipulate information on a larger and faster scale through generations of devices that decrease in size, weight, and cost. At the same time, the seeds of the next wave of innovation are sprouting through our crude, but accelerating, capacity to manipulate matter at the atomic scale in the ways we are currently manipulating information.

The following is by no means a comprehensive list, but rather a description of the progression of order through some important levels of power it has already achieved.

1. Order found the capability of manipulating order in the form of elemental matter when it found DNA. It did so through physical and chemical interactions in an increasingly orderly environment. This created the power of life on Earth.

2. Life then found the capability of manipulating order in the more orderly form of neural matter, through evolution. This created the power of reason.

3. Animals then found the capability of manipulating order in the form of existing matter (tools), through reason. This created the power of technology.

The Meaning Of Life

4. Humans then found the capability of manipulating order in the form of elemental matter, through technology. This created the power of chemistry.

5. Humans then found the capability of manipulating order in the form of computer memory, through chemistry. This created the power of computing.

6. Humans then found the capability of manipulating order in the form of biological matter, through computers. This created the power of genetic engineering.

      Through this progression we can see how order is finding higher order by the most convenient available means. Building on the powers it has currently, order finds more precise, purposeful, predictable, and complex ways of arranging matter to gain new powers, to find higher levels of order.
      As order's new capacities to manipulate matter reach maturity, we will have smaller and smaller devices entering every type of tissue in our bodies, and spreading throughout our environment, repairing disorder, and building new forms of order to our specifications.
      This new age of innovation will take the work out of everyday life, enabling our machines to sustain us as we pursue the things we find interesting. One day our manipulation of matter could reach a level of maturity that we can build a massive self-sustaining spaceship that absorbs and assimilates disorderly material as it travels through deep space. In a few of our lifetimes this spaceship could reach a nearby solar system with millions of the highest current carriers of order to take over its planets. Through this expansion, our order could even outlive the lifespan of our own sun.

Chapter 13. Is There A God?

As I said in the beginning of this book, I don't know. However, our understanding of order can shed some light on the nature of our creator should he/she exist.

The more we understand the biological mechanisms that comprise the order of a human being, the more apparent it becomes that we are basically fleshy robots. We are still a marvelously complex and beautiful wonder to behold, but every system in our body acts in a consistent mechanical way in concert with our other systems to complete a task. The difference between human beings and robots lie in our necessity and capacity to sustain ourselves.

Because we are the highest order beings, the robots that we create are an extension of us. Their tasks are performed to further our order. It's easier to make them perform their tasks while we supply their energy and repair. It would be much more difficult, but not impossible, to make them self-sustaining. If we did so, we would not see any magical soul within them, yet most of us see magic within ourselves.

We want to believe that there is magic within us. It's comforting to believe we will go to a better place after we die. When we are wronged we want to feel there is some higher system of justice that will punish our enemies, and reward us for all the selfless things we have done when no one is looking. We want a better understanding of our purpose and surroundings to guide us through the disorder in our lives.

These deep desires, common among all human beings, are responsible for the remarkable level of influential order present within the order of many of our most sacred texts. Organized religions have provided mankind with solutions, however imperfect, to all of these problems and more. They have a definite and undeniable power that is used for increasing order even though much disorder has also come about as a result.

The Meaning Of Life

    In the next chapter I'll discuss the system of society, which we have all adopted to varying degrees. The adoption of religion can give a person many powers, such as a mature system of society, while stunting other powers, such as reason. When we adopt a system of religion we may believe we are doing selfless acts in the name of our god/gods, when we are in fact acting to bring about higher order in our own system of religion.

    There are no truly selfless acts. Even when acting anonymously to help someone, we act through a mature system of society that assumes some responsibility for that person's order. We do so because we empathize with that person, taking on the disorder we perceive in them. We help them to remove that disorder in ourselves.

    Doing deeds out of empathy can bring us great order. When we see that order as the result of a religious system, we strengthen our belief in our god from the proof of the good he/she can bring about in us. When our actions strengthen our belief in our god we get the added benefit of reinforcing our trust in the god we depend upon to protect us from the disorder of death, injustice, and ignorance.

    When we act to bring higher order to our personal system of society by helping a stranger in need, we usually create higher order in those we help, and higher order in ourselves, creating a win-win scenario that leaves behind higher objective order in mankind. This is partly why the system of society is such a high order system.

    Many of us participate in philanthropy, giving some of our order to those we feel are most in need of it. Perhaps we give because we've reached a level of happiness in our lives that we want others to feel, and see charitable giving as a way to right some of the injustices in the world. A simulation of the world exists in each of us as a system of experiences in our minds. Those of us with mature systems of societies have empathized with many beings on the painful end of an injustice, assumed some of their disorder for ourselves, and

The Meaning Of Life

been unable to fix their disorder. Philanthropy helps us to do so, however indirectly, and through it we can receive much higher order than we sacrifice.

Much philanthropy is done through religious systems. Many people are exposed to the greatly beneficial philanthropic exchange of order through their religious organizations. Organized religion has persisted throughout the years partially because of the wisdom inherent in the sacred texts. However, there is a greater wisdom in the pursuit of higher order, for it is an all inclusive path of understanding that continually finds better ways to help mankind.

Religious texts exist primarily as a system of rules to guide our way of living. Successful religious texts gain influential order by being functional to those people that adopt the rules of the text. There are three main reasons why following an organized religion can lead someone to unintentionally create order at a slower pace, or even to create disorder.

Religious texts like any system of rules have to exist in a concise enough form to be recorded and copied. Therefore, they have to exist without all of their many exceptions. However, unlike all other rules, religious rules are given special weight as they are claiming divine origin. When we follow a system of rules as unquestionable doctrine, we are discouraged from breaking those rules to learn about their system. This method of living stunts the improvement of our order by limiting our ability to learn about the systems to which the rules apply. When we fail to learn about the systems of order around us we miss many opportunities to break rules according to their exceptions, which creates higher order than following them blindly.

Secondly, the system of rules is copied without variation from generation to generation. Any set of objectively functional "rules to live by" is tied to the environment in which those rules were written. Many of those rules need to be modified to stay objectively functional in the constantly

The Meaning Of Life

changing environment. Let me describe what I mean by "objectively functional":

A rule can be said to be objectively functional if its existence brings about higher objective order within the universe. In other words, a rule is objectively functional if it helps order's progression over disorder. Some objectively dysfunctional rules are created by a false approximation of the effects that rule will have on those who follow it. A rule may even have high functional order for the system to which the rule applies, but still have lower functional order for the universe, making it an objectively dysfunctional rule.

Some of the rules written in the ancient texts are still objectively functional today, while others are not. Because of our inability to predict the environment of the future, it is impossible to create rules that will be forever objectively functional. It's likewise impossible to know which rules are still objectively functional in today's environment. The United States constitution is almost certainly an objectively functional set of rules because of its built-in capacity to evolve. Through new laws and amendments we can modify our system of rules to reflect the changing systems that they govern. The United States constitution is far from perfect, but its high level of order is apparent by the rate at which carriers of humanity's highest levels of order flock to the country it governs.

The third main reason why following an organized religion can lead someone to unintentionally create order at a slower pace, or even to create disorder, is that organized religion perpetuates faith. Faith is the belief in something for which there is no proof.

Faith is an inherently disorderly concept. When we take things on a matter of faith we circumvent the natural process that has made us the carriers of the highest order. We almost always do so to avoid disorder, but the disorder we're avoiding is often disorder we should be facing and attacking with our ability to reason.

The Meaning Of Life

When we accept certain axioms or fundamental truths, we build order from those axioms in many of our systems. As we grow more order built on the foundation of these axioms, our ability to revisit and revise these axioms becomes more and more painful. To accept the fallibility of these axioms is to accept the fallibility of the order we have built upon them.

When we accept an axiom through evidence we can usually incorporate new evidence that illuminates an exception to that axiom, by only slightly modifying that axiom. These slightly modified axioms require little pain to incorporate into our systems. This is because that axiom was merely our best understanding based on past evidence. When we accept axioms through faith and we find evidence illuminating an exception to those axioms, either we accept our faith more loosely, deny the evidence, or build an increasingly convoluted system of order around the new evidence to preserve our faith. Either way, we are stunting our ability to understand the truer nature of things and complicating our ability to incorporate new knowledge in the future. To accept how some evidence is contrary to our faith would require us to doubt our whole system of security, often alienating friends and family.

The power of religion lies in its ability to calm all of our fears, to right all of the injustices leveled against us, and to answer all of our most difficult questions. Religion can take much of the pain out of living, but pain is the mind's recognition of disorder within a system that is important to it. That disorder exists and when we use religion to build order in place of that disorder, we are looking for the easy answer to cancel the pain rather than the correct answer to help us avoid it in the future.

When we look to religion to calm our fear, we put ourselves in God's hands. When we look to religion to right an injustice, we leave it to God to right the wrong. When we look to religion to answer a question, we trust the solution is a part of God's plan. Since we all agree we could never understand God's will, when we look to religion to calm these types of

The Meaning Of Life

disorder, we are accepting that we can't possibly understand and avoid an outcome we fear, that we can't possibly fix an unjust system, and that we can't possibly find the answer to our question.

Because religion offers such a potent elixir to our most intolerable pains, it has become a very important system. When so many people put their faith in a common set of principles, those principles assume great power. However, the advancement of order is always chipping away at the order contained in our unchanging ancient religious texts. The texts themselves stay the same, but our interpretations of them are constantly becoming less literal.

Churches are splintering into groups that are more tolerant of certain actions once deemed sinful to avoid the desertion of their members. Even denominations that claim to strictly follow the doctrine of their original texts ignore many passages as relics of the age in which they were written. Order is constantly defeating even the most powerful and omnipotent of pain relievers.

Perhaps the most painful part of life to find peace with is its end. Death was an important mechanism for life to find before it could truly flourish. If a life form was made to live forever its community would grow too large for its resources and its order would die off.

Our bodies have a life cycle whereby the cellular maintenance of our order atrophies with age, yet our knowledge and wisdom increases. However, in most cases we become entrenched in our beliefs and in doing so we lose hope for improvement from the status quo. Death is constantly creating space for higher order to flourish. New power structures, new ideas, and new capabilities emerge from a fresh look at old assumptions.

It's painful to understand life without a soul because death is inevitable. We rely upon the order of our body and the order of those we love. When that order is lost we crave an understanding where that order will one day be restored. Most

The Meaning Of Life

of us choose to live with such an understanding for the comfort it provides, and in doing so we don't fully recognize the fleeting gift of time.

There is still comfort in an understanding of order in that the actions we take and the order we produce will live on as long as the order of life spawned from Earth progresses. Future generations will stand on the shoulders of past giants as we do today.

For some this is sufficient comfort. Others will make a potent counterargument; "Why not believe in a soul for the relief it provides from the pains of life? Is it wrong to care more for the sufficient order of human comfort, than the order found from rejecting those comforts in a frenzied pursuit of some higher order? If the human race controls the speed of order's evolution, why not throttle it at a comfortable pace?"

To answer these questions we should first understand that no single philosophy, worldview, or religion has a monopoly on comfort. An atheist may struggle with the gravity of death while a Christian struggles with the guilt of sin.

A Muslim, who doesn't believe in the concept of order vs. disorder, feels pain just the same whether or not he/she understands it as the mind's recognition of disorder in a system it deems important.

There are many ways to deal with a problem. Religion gives us a way to relieve the symptoms of some mental stresses and some rules to follow that will often help us avoid pain in the future. An understanding of order provides less relief from the symptoms of some mental stresses but great surgical precision in learning how to avoid them in the future.

Sometimes addressing our symptoms is sufficient to help us concentrate on producing higher order in areas separate from our source of pain. Through this mode of living we can still create the highest levels of order within our individual specialties.

The Meaning Of Life

On the other hand, a deep understanding of order can provide a lasting state of enlightenment, which combines a curiosity of all kinds of order, known and unknown, with a path towards discovering their source.

While an understanding of order can expose the flaws in any organized religion, it cannot provide insight into the existence of a God. Somehow our universe began. Somehow the initial preconditions of matter, energy, space, time, and the four fundamental forces came into existence with specific properties that have initiated order's consumption of disorder. This great mystery is ill served by every form of organized religion, and unanswerable even through the pursuit of higher order. This question remains, however futile, most interesting to ponder.

Chapter 14: The System Of Society

A system of society exists within all of us. We have all received order from cause and effect interactions with other forms of life. Our systems of society are intricately tied to many of our other systems but can still be isolated through our shared understanding of life and empathy. One person's system of society can vary greatly from another's both in its scope, and in that person's capacity for empathy.

Man has evolved the capacity to empathize with other forms of life. When we think about society we generally reserve the term for our fellow human beings. Many of us have developed a broader scope of life forms that we feel empathy for, and our experience with all of these forms comprise what I refer to as a person's "system of society".

When we recognize the signs of pain in other life forms, we try to understand their pain by empathizing with them. The more we care about their order, the more painful their pain is to us. When we empathize with other beings that we care about, our perception of their disorder becomes our disorder. When we act to relieve some being's pain, we are acting to relieve our own pain.

We have adopted our systems of society because we have needs that are best met by other people. We can learn from their successes and mistakes or in some way benefit from the order they can supply. So, to better understand another being so we can extract the order we desire, we map our systems to their similar systems and approximate the effects of our differences. As a means to get the order we need for ourselves, we have taken on the responsibility of helping to supply the order others need. In doing so, we build up a reserve of goodwill which acts like informal money that we can exchange for something we need. Just like exchanging money, our exchanges of goodwill provide us all access to win-win scenarios.

The Meaning Of Life

Our recognition of the benefits of our systems of society has provided us with an extremely functional resource for furthering our own order. Through helping others we not only benefit from the win-win scenario by exchanging some order for higher order in the form of goodwill, but we also better order our systems of society by improving the order within the members of those systems. This provides us with an opportunity to participate in the ever-present but more elusive win-win-win scenario.

In many cases, we will never again meet the stranger that we helped, to cash in on their goodwill personally, but our generosity will often create a desire by that stranger to do good for another stranger, increasing the order of the universe. In these cases we get the satisfaction of bringing order to our system of society, the stranger gets the order he/she needs, and some other stranger gets the order he/she needs.

As we start to develop a system of society, we care most about maintaining the order of the members of our system of society that commonly supply us with the order we need. In the typical case this is our parents. They give us food, shelter, knowledge, and love, as we grow capable of finding these things for ourselves.

We first build empathy for members of our family because they are most likely to reward us for any order we give them, therefore the benefits of this order exchange are the most obvious to us. Soon our systems of society can evolve to include friends, co-workers, countrymen, humankind, and other animals (not necessarily in that order). It's easier to empathize with a being that is familiar to us. It takes less effort to exchange order with beings that share our experiences as well as our spoken, written, and body languages.

As our systems of society evolve, we better understand our unwillingness to empathize with beings that we find strange as a symptom of intellectual laziness. Some of us maintain a need to stay curious, to nourish our minds with the higher order that's only attainable through reaching out to more and more alien groups, cultures, and species. This curiosity can be found from the understanding that beings similar to us will likely contain order that is similar to ours and therefore redundant.

We can choose a life where we work just to maintain a level of order that we have found to be comfortable, or we can indulge our curiosity, seeking out the disorder of change to build higher order in its place.

Many people shy away from discussing politics and religion because they expose the most basic precepts of our order. When we expose them we invite criticism of these principles. To change these basic principles would be painful because we would suddenly realize disorder in all of these other systems built upon these basic principles.

Therefore, when we talk about politics and religion we test our own willingness to face uncomfortable choices between painful incorporation of higher order wisdom, and stubborn rejection of wisdom for the lazy bliss of ignorance.

As we accumulate order we go from thinking we'll never understand enough to be successful, to achieving some success, to thinking we know it all. Many of us stop there. If we keep looking we begin to understand how little we actually do know and how fruitful the pursuit of higher order can be.

As our technology increasingly provides us with ways for functional order to gain influential order, generations of humans will break down the barriers that have separated us. The jet engine, the internet, and language translation software are just a few of the ways order is bringing humankind together through technology. This new interconnectedness works in many ways to further the progression of our order by exposing us to other cultures, building empathy and understanding between alien nations.

Chapter 15: Groups

Throughout time life forms have recognized the utility of forming into groups. When a group of beings form together for a common purpose they are better able to reach that common purpose.

Groups that operate through the direction of a single leader, when there is sufficient time to make a decision through compromise, are more prone to act against the interests of their members. When there isn't sufficient time to reach a compromise, groups that follow a single leader are prone to act more functionally. Groups that operate through consensus and groups that operate through the direction of one leader will likely still be more functional than the members of those groups acting alone. If the size and maturity of a group is sufficient to share power between a deliberative body and an executive body, that group will prove even more functional.

Because most of us have recognized this, we become members of a group and turn over some of our decision-making powers to the decision-making body of that group. The decision-making body may then set policy that governs the actions of the members of that group.

As individuals acting as a part of these groups, it can be difficult to see how every action we take is for the purpose of furthering our own personal order. Perhaps we're in a job with an angry boss that orders us to do useless work. If we believed we could get a better job and manage the lack of income while we search for it, we would. Perhaps we are oppressed by a militant government that starves us and threatens our lives. If we thought we could escape to a better place with our lives and perhaps the lives of our family, we would.

Sometimes we are too afraid of disorderly uncertainty, too set in our ways, or too skeptical of our own capacity to bring about the change we need. Sometimes we feel responsible for the order of our friends and family that rely

The Meaning Of Life

upon our income or attention. Groups that have evolved enough order to offer us something worthy of membership have most likely evolved many other mechanisms for retaining the members of their group. Whether a group offers its members carrots or sticks for staying with that group, the members who stay do so precisely to further their own personal order.

A country is a special kind of group. It has a decision-making body, a government, with ultimate authority over its defined territory. Governments are responsible for more than just setting policy. Mature governments understand their responsibility to maintain a system of justice. This involves adjudicating disputes, defining criminality through law, and punishing violators among other things.

A system of justice gives all members a set of rules to abide by in return for a reasonable expectation that others will also abide by those rules. Justice can be said to be the preservation of objective order. Just as we are unable to perfectly determine the objective order of anything, we are unable to perfectly determine what is just or fair.

The preservation of order sometimes involves the destruction of order in a person who has destroyed the order of another. When a government does so it maintains the trust of its members in that system of justice.

Because we are unable to determine the objective order of any person, most mature systems of justice strive to treat every person equally. This concept of equality arose to defeat the injustices of racism, sexism, and other prejudices present in the ruling classes. Because everyone has an objective level of order, one might mistakenly assume true justice would best be served by giving preferential treatment to those we approximate as having higher objective order. Any mature system of justice has to operate blind of any approximations of objective order. Doing so provides a level playing field for all carriers of order to achieve higher order free of some oppression from mankind's imperfect approximation of order.

The Meaning Of Life

So our mature systems of justice approximate the most just solution to disputes taking into account the rules, as they were when the alleged injustice occurred. They do so on the grounds that every person is equal. The system of laws is given greater order through legislation when a decision-making body approximates a just way to do so.

As our systems of justice evolve, so does the order of the members of those systems. Not only do the members gain trust in their justice systems, but the environment those members exist within becomes more orderly.

Chapter 16: Free Will

Try to think of one choice you've made in the last 24 hours where you truly don't know why you made it. You might find something that you would have done differently, now knowing the outcome. That just means you're learning, and not perfect. Days after a choice we can look back confused as to why we made that choice, but only because we have lost the memory of the state of our systems as they were when we made the choice. We can even look back on a choice and think "I knew better than that", but we in fact didn't or we would have acted differently. Sometimes we need reminders of the importance of some of the rules we've established for ourselves, yet continue to break.

      A person, trying to understand free will, could throw away some money in an attempt to prove it's possible to act with free will against his/her order. The assumption from this proof would be false however. By throwing away money for this purpose, he/she has accepted that the loss of money is worth the order gained by reinforcing his/her disbelief in order.

      I challenge anyone to find a choice they've made that wasn't what appeared to be the best choice at the time. We may have an array of choices presented to us, but we don't make a choice, we approximate which choice better furthers our order, and then that choice is the only choice to choose, which is the same as having no choice.

      We don't make choices, we make approximations. As we're making these approximations, our subconscious mind is doing most of the work simulating outcomes from an action, approximating the increase in order to our systems that action will bring, providing us quickly with an answer and justifications for that answer. These answers usually come in the form of a "feeling", which is an approximation without hard numbers. Because our brains are not equipped, like computers are, to compute their answers starting from exact numbers with known precision through known algorithms to an exact solution, we sometimes experience these approximations as a somewhat magical "choice" that came from our soul.

Because we have such large complex systems in our brains, we cannot understand perfectly how our approximations arise. We have tasks that we execute to bring about higher order in ourselves, and the universe. We don't have to know exactly why a certain task is so effective as long as it has proven reliable.

A brain has limited capacity. It takes most of the capacity of our brains just to contain our systems. To contain our systems, and a thorough understanding of all of those systems would take so much more capacity. This extra capacity would be wasted if spent trying to understand perfectly, systems that work well enough as they are. Therefore, it has proven more orderly to operate without a perfect understanding of our approximations. We shouldn't misinterpret our imperfect approximations as being the active hand of a spirit, but the necessary result of order acting through beings with limited neural capacity. We see psychologists and/or look critically at ourselves to try to better understand our systems of need.

Our actions are determined by the functional order of our minds approximating the most orderly path to guide our bodies. Order is taking over disorder through every action we take. Order doesn't know the best way to find higher order, just as we don't. Order searches for higher order through our approximations and actions. The search for higher order is the only thing that guides us, determining our actions. This does not mean, however, that we don't possess free will. In fact, the reason the question as to the existence of free will has proven so perplexing for philosophers is that the answer is both that it does and does not exist. The question of free will can be defined as "whether rational agents exercise control over their actions and decisions."

The Meaning Of Life

    As carriers of order, we can see ourselves as isolated, owning the space within the boundaries of our skin. We are, however, systems of matter that are arranged as a result of how matter, space, and time began and how the fundamental forces have acted upon them since. Our actions are guided by our internal circuitry and our sensing of our environment. When we choose to act a certain way, we've actually approximated that way on our own. Therefore, you could say we have exercised control over our actions and decisions when approximating the best choice.

    This is true, but not the whole truth. The explosion that created our universe set the initial conditions that determined how all matter would act from then on throughout our universe. We are a part of that matter and though our available choices are determined by our environment and knowledge of our choices, our minds are solely responsible for the approximations that determine our actions without any active spiritual influence. However, the arrangement of neural matter that makes up our minds is a result of those initial conditions of our universe. Therefore, those initial conditions, be they set by a God or not, determined our order and our approximations. We do not have free will to act against the destiny laid out by those initial conditions.

    Free will is only a property of a person when you isolate that person from the environment. Because we are all created from, sustained through, and will return to, our environment, our isolation from our environment is artificial.

    Free will remains a useful concept to use when trying to understand the self, and as a concept it exists as order. We cannot know the path the matter of the entire universe will take, so we artificially isolate ourselves from all of the matter beyond our skin to analyze properties of the self.

    When we manipulate our environment to make it better suit our needs we can be seen as isolated from it because *we* are manipulating *it*. However, from someone else's point of view we are a part of their environment, so when we

The Meaning Of Life

manipulate our environment, their environment is manipulating itself. We can see anyone as isolated from the environment or as a part of it. Any time someone manipulates their environment, they are playing their part in the progression of order of the system of all matter.

Our actions come from our minds, and our minds come from the progression of order throughout the universe. To say our actions only come from the self is to deny the conception of the self. To say they only come from the universe is to deny the functionality of isolation.

Through this understanding we can hold people accountable for their actions for the functional order learned through consequence, while still finding compassion in their inability to have taken any other path.

We should understand that we are responsible for our actions and for learning from our mistakes. We are more functional when we do so. It may be therapeutic to recognize that we had no other choice, having done the best we knew how, but that shouldn't serve as an excuse for complacency in our future actions or current preparation. To use this knowledge in such a way would create disorder within our systems that approximate the most orderly path.

To think that because our eventual actions are already determined for us, we don't have to do the necessary hard work to determine our best action, is to put this knowledge in a disorderly context. If we see it this way then our eventual action, that was predetermined, will have been predetermined to be the action of a person that is unprepared.

The Meaning Of Life

Chapter 17: Everything

I chose the title "The Meaning Of Life" with full recognition of the brazen audacity in laying claim to such an understanding. It was only through further contemplation on the nature of order vs. disorder that I came to see this title as somewhat humble. The battle of order vs. disorder determines the progression of everything in the universe.

The long life of a carbon atom is a fascinating thing to ponder. Born in an ancient star, indistinguishable from all other carbon atoms, cast off to the far depths of space from the powerful explosion of a supernova. It slowly feels the gravitational pull of a distant gathering cloud of matter that forms and compresses into planet Earth.

This atom bubbles through the cauldrons of magma beneath the cooling crust of the planet for billions of years until launched into the sky from an ancient volcano. Grabbing onto two oxygen atoms it swirls for months above the Earth until wafting past the leaf of an early plant.

Sucked into the gears of some common plant matter, this carbon atom is separated from its oxygen atoms and guided through the mechanics of photosynthesis where it is deposited as part of a branch.

Once the plant has died and decayed, the carbon atom is whisked around the Earth's surface and resettled by rainfall and evaporation cycles. Again it finds two oxygen atoms and takes flight starting the process over again. Eventually it becomes a part of some plant matter that decays into some soil with the right composition to grab onto it tightly enough to keep it within that soil for thousands of years. It stays in a location where layers of soil are deposited on top of it, burying it deep underground, causing it to pressurize and become a part of an oil molecule.

One day it is pumped out of the ground and separated from its neighbors into gasoline where, through combustion,

The Meaning Of Life

the energy that bound it tightly to its neighboring atoms is converted into mechanical energy to push someone's car forward, releasing it once again into the atmosphere. Again it finds a plant and soon becomes a part of some man's meal. The man eats the plant and the atom becomes a part of him.

Here the atom has found its most important role yet. It is now part of some machinery that helps to fulfill the wishes of the mind of a human being. Being part of a life form, it is just a temporary placeholder that will, at some point, be swapped out to equal effect by another atom of the same type. However, this is the carbon atom's time to fulfill its role, and it does so perfectly.

Perhaps it was part of the arm that didn't throw the ball quite fast enough to evade the batter in a baseball game. It played its part exactly as any other carbon atom would have, the result of which helped the pitcher to understand his need for further practice.

A carbon atom may be interchangeable with any other carbon atom to equal effect on the order of the man, but a human being is not interchangeable with another human being to equal effect. As the building blocks of matter form larger and more complex structures, from a carbon atom to a carbon dioxide molecule to a protein used to build muscle fiber, to a human being, the structures become imperfect.

These imperfections are the necessary variations that make each human a unique experiment. Because of these variations we can succeed where no others have succeeded, or we can fail.

We are order's best mechanism for finding higher order and this is our great importance. It's a common human experience to ask the great question; "What should I do with my life?" This leads us to wonder about the meaning of life itself. To understand this we have to step back from our recent personal experiences, our fears, our emotions, and even our deeply held beliefs. We have to float above ourselves and see the forest apart from the trees that usually fill our view.

The Meaning Of Life

We have to find a context in which everything fits. The underlying mechanics of what is happening on the smallest levels to make one atom can remain a mystery so long as we know that one atom behaves like every other atom of the same type.

This is where order is spread. Successful functional order gains influential order by copying itself or being copied. Perfect copies are made of the same types of elements in the same configuration. Many copies are not perfect, but are close enough to exhibit the same behavior. It's not the actual atoms that make the order, just as it's not our tissue that makes our order. The individual atoms will be exchanged throughout our lives as we consume new food and replenish our structures. Our great order comes from the complex arrangements of those atoms. These arrangements are the important essence of our beings.

Imagine this thought experiment: One day we could potentially scan a human being, precisely identifying the location and elemental type of every atom. We could then create an exact copy of that person by placing another one of every atom that constitutes that person in the exact same relative position a few feet away.

Yes, we would have to pass many extremely difficult technological hurdles to make this work, but the practicality of this is not important to the greater point. What we would have created would not be merely an identical twin or a clone, with the same DNA, born at a different time. Instead, we would have created an exact copy.

This new human being would be exactly the same as the scanned human being with all of the same experiences in only a slightly different context. Both beings would respond exactly the same to the same questions or stimulus. This simple thought experiment leaves no room for a metaphysical soul, unless a soul can be copied or created by man. Perhaps it could be believed that the soul resides in the original man. So what if instead, the machine dissected a man and made a copy

The Meaning Of Life

of each side to connect with its original counterpart. You would end up with the same result; two guys that both think they're the original, neither of which containing a unique soul.

It is true that one atom is not exactly like another atom of the same type, yet every atom of the same type will still function exactly the same as a piece of some system of order. An electron can orbit the nucleus of an atom in a unique direction from all other electrons, still when its atom replaces an atom of the same type, it functions the same as that atom. It occupies essentially the same space with essentially the same angle of connection to its neighboring atoms. In doing so, it serves its role in completing the order of its system in, not just essentially, but exactly the same way.

There are many examples of this phenomenon, such as the transistor. Two transistors can be made imperfectly with some extra atoms here or there. When we manufacture a transistor we have imperfect but acceptable methods of producing a transistor. As long as the output voltage, response time, leakage, and a few other characteristics of the transistor are determined to be within acceptable ranges of operation, the transistor will function exactly the same as other transistors in its circuit. It is, therefore, perfectly interchangeable with another transistor in its circuit.

This phenomenon was also described in the chapter on context when discussing the mosaic parable. If we were to rotate the arrow mosaic that points to the right by one degree clockwise or counterclockwise, the tourists would still take the same path. Similarly, an atom can vibrate around in its place due to heat so long as the vibration isn't strong enough to break its bond with its neighbors, changing the order of the system to which it belonged. An atom is either bonded to it's neighboring atoms or it isn't, there's no in between. Once that bond is broken, the order of the system to which it used to be bonded to changes. The newly freed atom now has a different level of order as a gas or a liquid. In a solid state, an atom can hold information as a placeholder in a piece of order that may

The Meaning Of Life

or may not be passed on. As a liquid or gas an atom loses the property of maintaining information by existing in a fluid state, but it still has a level of primitive order, which is limited to functional order. We can create order from a fluid or gas that can gain influence by, for instance, painting a picture of the sea that becomes popular. However, the sea required our ability to freeze a picture of it in our mind or in a camera through our technology to preserve the order of that fluid. The fluid could not preserve its order alone.

A molecule of carbon dioxide has a low level of order that can be easily copied. Our order is much more complex, and therefore difficult to copy, but if it were copied, the copy would be no different from the original at the moment it was copied. The point here is that a human is an arrangement of atoms, and that arrangement is what is important and unique about any human, not the atoms themselves.

For instance, a person can become anemic from an iron deficiency, but the lack of iron atoms in the body is just the symptom of a problem with the person's order. That person's order causes him/her to use more iron than he/she intakes. You could sneak an iron pill into that person's meal, and the symptoms would go away for a while and then reappear. It's only when the arrangement is corrected, for example by the person learning the source of his/her anemia and that he/she must take iron supplements, that the problem will go away.

Of course matter is a crucial prerequisite for order. It is the medium in which order exists. But the arrangements are where the level of order is found. Order plays by the rules set by the properties of matter, and finds ways within those rules to increase its power. This explains the shape of every leaf, the sound of every laugh, the taste of every fruit. An understanding of order explains more than just the meaning of life. It provides a path towards understanding almost everything.

The Meaning Of Life

Chapter 18: The History Of Order

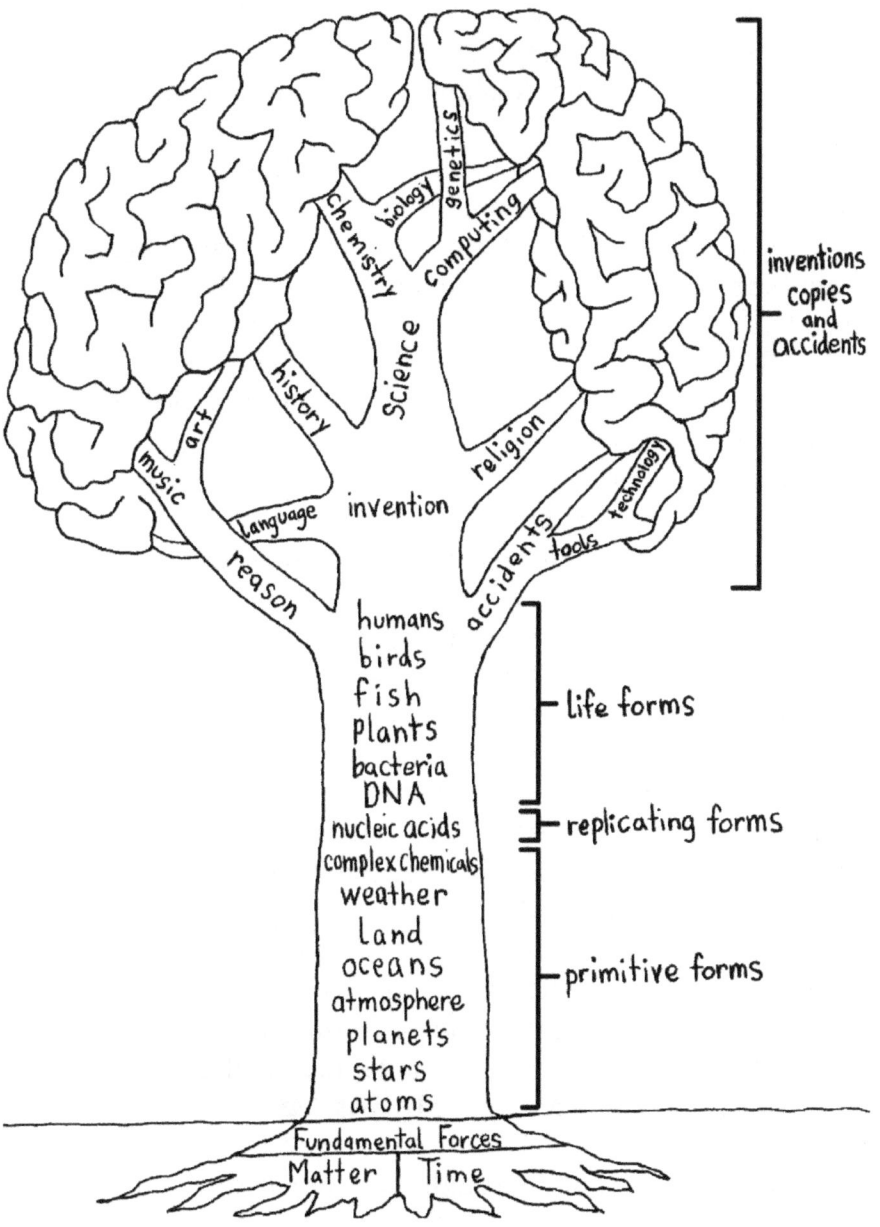

The Meaning Of Life

To reach an understanding of the role humankind plays in the progression of order, it is useful to visualize the history of the progression of order to today. The illustration of the tree of progress is a crude attempt to show order's advance through the ages.

In it, the fundamental medium through which order grows is shown below the ground as the root system. At the time of the Big Bang, all matter existed in a state of change through time. The manner in which it has changed has been defined by the four fundamental forces ever since.

After the Big Bang sparse clouds of helium and hydrogen matter slowly coalesced through the attraction of gravity, forming stars. The extreme pressure from this compaction caused nuclear reactions that formed heavier elements such as carbon. This new matter was then expelled through supernovae explosions that later compacted into planets such as Earth.

As these planets formed, the Earth was left with an atmosphere where water could condense and fall as rain. This cooled the molten surface and created an ocean and a cooling crust.

On the land, structures could keep their primitive order for long periods of time yet there was still a constant rearranging of order through shifting tectonic plates, weather, and other powerful forces of nature.

These constant sources of disorder were necessary for new combinations of matter to continue forming. New chemical mixtures were always being formed as the elements of the Earth's crust were absorbed and mixed in the surrounding water that was moved by wind, evaporation, and tidal motion.

New order is found through these processes. It may happen slowly, but there was no shortage of time. After hundreds of millions of years these planet-wide atomic scale experiments likely created the first replicating molecule. It

The Meaning Of Life

may not have been as structurally sound as some of the rocks surrounding it, but the order of any one instance of this arrangement didn't need to survive a wave of disorder as long as it had replicated its order sufficient to have a copy of itself far enough away from the disorder to outlast it.

At this point order had found redundancy. This redundant order quickly spread until likely hundreds of millions of years later one of these replicating molecules found the order of variation, which created life.

Now order had found that an even more effective way of preserving order was not to replicate itself perfectly, but to slightly vary each replication. In doing so, the nature of competition was created and survival of the fittest would form the new paradigm of order's advancement.

Bacteria grew and spread past numerous barriers to its existence. It found ways to exist on a wide range of diets and surfaces. It even found ways to compete with other forms of bacteria. Soon it found the order called flagella, where the cell of some bacteria could propel itself to find resources rather than waiting for its environment to bring it in contact with the resources it needed.

In time these single celled structures advanced into multi-celled organisms. As the structures became more complex, organisms found the order of sensory organs with conductive connections between these organs and the structures that drive the organisms. These conductive connections found arrangements that helped the organisms to avoid common types of disorder through instinctual reactions to sensory input.

Out of these conductive connections, the first neural tissue was formed. This neural tissue, in the crudest of ways, could learn from it's own mistakes and improve it's behavior. The success of these life forms was no longer determined simply by the nature of their genes, but also by the order they gained through experience.

The Meaning Of Life

At that point, order had found a much more efficient way to advance itself against disorder. The powerful organ of the brain began to flourish through the animal kingdom. Plants had no need or capacity to evolve a brain because they were rooted in the soil. They needed only to adapt well to their soil and the nutrients common in their location for their order to persist.

Animals were the wanderers that had to learn how to adapt to a variety of environments to successfully scavenge and compete with other animals for the resources they needed. Fish, birds, and many other creatures lived out the experiments of their order in fierce competition to stay alive and pass on their order.

From the order of neural tissue came the power of reason, which is in itself a form of order. Reason is an arrangement of neurons in sufficient order to take input, compare it to experience, and generate an approximation. We use this order to approximate how best to manipulate our environment to further our order, as did fish and birds.

At a point in time, before order had found human life, a bird holding a stick in its beak poked the stick into a hole in a tree and pulled out a grub. At that time the bird learned how to use arrangements outside of its body to help it find the order it needed. Because it could reason, it recognized that the stick helped it to find the grub and this was one of the earliest ways in which order found technology.

The use of tools progressed rapidly with the order of the human being. Early man found the capacity to not only find tools, but to create them methodically out of the primitive order around them.

Technology would spread and grow in a variety of fields as the order of the human brain brought exponential growth to the present levels of order on Earth. Humans expanded on the tonal languages of bird songs and the body languages of apes, creating characters that could be arranged into words and recorded.

The Meaning Of Life

Our spoken language traditions spread some functional lessons through the ages. The power of written languages helped us to pass on similar lessons with more accuracy.

Soon our thirst for wisdom produced existential questions with no satisfying answers. A need was created for an all-encompassing philosophy to guide our actions. The power of the written language was harnessed by ancient wise men, and religions that had once been passed sloppily through spoken folklore and superstitions were now coalescing into literary works that explain everything as a part of some god or gods' plan.

At around the same time, the seeds of the order of science were planted. Mankind found a consistent way to search for answers to the difficult questions through analysis of evidence.

The progression of science greatly furthered the order of our technology into the realms of chemistry, computing, and genetic engineering, to name just a few.

Looking at the illustration of the tree of progress, it's easy to get stuck at the level of order of the human and wonder why the next levels are of abstract things such as language, history, and science.

To understand order, it is important to understand that this is not at all a leap. We may experience the order of a fish, a bird, and a human completely differently from that of the concept of science, but that's only because they happen outside of our minds whereas the concepts of science happen within our minds.

Science, like a life form, is an arrangement of matter that takes an input and produces an orderly output. It, like language, religion, and all other concepts of the mind, is a function made of matter that serves the purpose of furthering order.

When biological order reached the level of humans, it found all the capacity it needed to reach higher order in the mind of the human. The slow evolution of genetic order continues to progress with every generation, but its changes become decreasingly important as the level of order within our memes evolve at an exponential pace.

Now, our neurological order is still advancing with us humans as willing hosts and beneficiaries to its progression. It will continue to do so as we incorporate our technology with our biology creating even more orderly hosts.

The idea that order's search for more resistance to disorder guides our every action is beautiful in its simplicity. However, once we look deeper into what determines the level of order of a system, how arrangements have found such clever methods for resisting disorder, we can marvel at the complex forms that a universe full of matter, with billions of years to interact, can produce.

The universe is full of mystery, much of it contained within our tiny precious Earth. Mankind yields the greatest known powers in existence. Wonderment and insight can be found from the careful study of a simple leaf if we just look, ask questions, and seek out their answers.

Every human being has the potential for greatness. We need only to appreciate our uniquely high order, and make the most of the time we have, to find joy in our contributions to the progression of the universe.

The Meaning Of Life

www.ingramcontent.com/pod-product-compliance
Lightning Source LLC
Chambersburg PA
CBHW032148040426
42449CB00005B/441